THE WHICH? BOOK OF
PLUMBING
AND CENTRAL HEATING

THE WHICH? BOOK OF
PLUMBING
AND CENTRAL HEATING

DAVID HOLLOWAY

Which?
BOOKS

Published by Consumers' Association
& Hodder and Stoughton

THE WHICH? BOOK OF
PLUMBING AND CENTRAL HEATING

Which? Books are commissioned by
The Association for Consumer Research
and published by Consumers' Association,
2 Marylebone Road, London NW1 4DX,
and Hodder & Stoughton,
47 Bedford Square, London WC1B 3DP

First edition 1985
First paperback edition September 1990
Revised reprint January 1992
Copyright © 1992 Consumers' Association

British Library Cataloguing in Publication Data
A catalogue record for this book is available from the British Library.

ISBN 0-340-57047-4

Thanks for choosing this book...

If you find it useful, we'd like to hear from you. Even if it
doesn't live up to your expectations or do the job you were
expecting, we'd still like to know. Then we can take your
comments into account when preparing similar titles or,
indeed, the next edition of the book. Address your letter to
the Publishing Manager at Consumers' Association,
FREEPOST, 2 Marylebone Road, London NW1 4DX.
We look forward to hearing from you.

Typeset in Great Britain by
Bookworm Typesetting, Manchester
Printed and bound in Great Britain by
Scotprint Ltd,
Musselburgh, Scotland

Acknowledgements

Design
Val Fox, Fox Design

Drawings
Tom Cross

Cover illustration
Peter Harper

Cover design
Philip Mann (ACE Ltd)

The author wishes to acknowledge the assistance given by various
manufacturers in providing information on their products
and the help given by the
Which? Product Research Group
in making invaluable comments and suggestions.

Particular thanks are due to John Love, MCIBS M Inst R, heating
consultant, for his unstinting help in checking the manuscript
and proofs and his major contribution to the section on central
heating. Thanks are also due to Jill Thomas and Bob Tatters all for
their valuable contributions.

CONTENTS

CONTENTS

PART TWO: Plumbing Room by Room

PART THREE: Central Heating

INTRODUCTION

The Which? Book of Plumbing and Central Heating is intended for the home owner who would like to carry out the majority of plumbing jobs in the home but lacks the necessary knowledge and experience.

It is not intended as a guide to the relative merits of competing products – *Which?* magazine, with its comparative testing reports, fills that role – but rather as a guide to the tools, materials and techniques needed for home plumbing and to the various plumbing jobs around the home, including the installation of central heating.

Plumbing undoubtedly involves a number of skills, but with the disappearance of lead as an important plumbing material and the introduction of easy-to-use materials and do-it-yourself 'kits' for plumbing fittings, it has become a lot easier and more within the reach of the moderately competent amateur. The growth of mail-order suppliers (particularly for central heating) and of do-it-yourself 'superstores' has made buying a less daunting task – no longer need the amateur feel embarrassed about not knowing exactly what he or she is looking for.

Plumbing can be physically hard work. Although many tasks need a degree of precision rather than one of brawn, there are also many (such as undoing reluctant nuts, making large holes in walls, removing old plumbing fittings, lifting floorboards and so on) where strength and not minding getting dirty are important.

The book is divided into three sections. Part One (Chapters 1 to 6) looks at the tools needed for plumbing and then at the various pipes and fittings used for the majority of jobs. Then the four main systems within a house – cold water supply, hot water supply, wastes and drains and rainwater disposal – are dealt with in turn. The jobs necessary to maintain or replace the important parts of these systems are all covered, many with step-by-step illustrated guides.

The second part of the book looks room-by-room at some of the plumbing jobs you might want to do: in the kitchen, in the bathroom, in other rooms and, finally, outside in the garden. Again, there are step-by-step drawings in addition to the main text.

The third part of the book covers the selection, design and installation of a smallbore central heating system. While not pretending to be a comprehensive guide to all the aspects of central heating, this section should enable you to save considerable sums by installing your own central heating in all but the most complicated of cases.

To find out more about the different makes of plumbing fittings, you really need to get hold of the manufacturers' catalogues. Not only do these give details of the various products but they also often have information about how and where the products are to be used as well as tips and hints on plumbing design and techniques. Addresses are on page 156.

In general, brand names have been excluded from the text, except where products are unique or where instructions are specific to a particular manufacturer's product.

Rules and Regulations

In many ways, Britain is unique in allowing householders to carry out plumbing (and electrical) repairs and alterations in their own houses. In many other countries, one or other (or both) of these is against the law.

However, there are many rules and regulations in Britain which it is advisable to follow – indeed, you are breaking the law if you *don't* follow some of them.

The main rules and regulations which affect the home plumber are:
- [ ] Water authority bye-laws
- [ ] Building Regulations
- [ ] Gas Safety Regulations
- [ ] Wiring Regulations.

Water bye-laws
The water industry in Britain is divided up into a number of autonomous water companies and councils (Chapter 3 gives details) and each is responsible for producing its own water bye-laws.

Although there are some fairly minor local variations (because of the quality of the local water, for example), most water companies base their bye-laws on a set of 'Model Water Bye-laws', so there is a degree of consistency across the country. Copies of water bye-laws are available (usually free) from your local water company.

The purpose of water bye-laws (made under the Water Acts) is to prevent 'waste, undue consumption, misuse or contamination' of water. The kinds of thing which the bye-laws cover are: pipes and pipe fittings, taps, valves, storage cisterns, hot water equipment, bathroom equipment and flushing cisterns. The bye-laws specify how and where such equipment may be connected to the household plumbing system and the types of fitting and system allowed. A particular concern is *back-siphonage* – the contamination of the mains supply with dirty water or water from a storage cistern being drawn back into it – and many of the bye-laws are designed to prevent this.

You are obliged by law to inform your local water company in writing at least five working days before making certain alterations or additions to your plumbing system; there would be stiff penalties if you were to be convicted of a contravention of the water bye-laws. An Inspector from the local water company may well visit your house if you intend making extensive alterations or additions and may be able to help you by spotting any bad practice or likely contravention of the bye-laws.

Many plumbing fittings for use in the home are approved by the Water Research Centre under its voluntary scheme – look for this approval (or a British Standard Kitemark or an approval mark to a relevant EC Directive) when buying.

The Model Water Bye-laws were extensively revised in 1986 and new local bye-laws, based on the Model, were introduced in January 1989.

The main changes are:

□ allowing for the first time storage-type unvented hot water systems, which are common in North America and on the Continent (see Chapter 4); unvented sealed-type central heating systems are already allowed
□ forbidding the use of capillary fittings containing lead-based solder on pipes supplying drinking water (see Chapter 2)
□ changing the size of flushing cisterns used with WCs (to come into effect in a few years' time – see Chapter 8)
□ prohibiting the connection of garden hosepipes to taps unless there are back-siphonage protection devices installed at the taps (see Chapter 10)
□ introducing new requirements for ballvalves, cisterns and frost protection (see Chapter 3).

Building Regulations

Although Building Regulations are mainly concerned with the construction of houses – and proposed changes in construction – there are ways in which they affect the home plumber, too. The relevant requirements (which have the force of law) are to do with **drainage** and **ventilation** of rooms containing WCs.

The Building Regulations themselves do not make for easy reading, but there are various guides designed to explain them and the DoE now produce a *Manual to the Building Regulations*. The Regulations are presently the responsibility of Building Control Officers within local authorities and it is to them that you should apply if you want to do any work which might be affected by the Regulations.

New Building Regulations were introduced in 1985. The Regulations themselves set out only the main performance criteria with the technical detail set out in supporting documents. Scotland and Northern Ireland have their own Building Regulations.

Gas Safety Regulations

By far the most important provision of the *Gas Safety (Installation and Use) Regulations* for the home plumber is that it is illegal for anyone to carry out any work in relation to a gas fitting who is not 'competent' to do so.

In practice, this means that you would be extremely unwise to attempt your own gas fitting. Although many of the techniques are similar to those needed for cold and hot water plumbing, the consequences of making a mistake are potentially much more serious and if you are found out, you could be heavily fined.

Qualified gas fitters are on the register of CORGI (Confederation for the Registration of Gas Installers).

Wiring Regulations

Except in Scotland (where electrical wiring is part of the Scottish Building Regulations), there are no legal restrictions on anyone carrying out his or her own electrical repairs and alterations.

However, local Electricity Boards have the right to inspect and test any electrical work they think may be unsafe and also to refuse a supply.

The electricians' 'bible' is the IEE Wiring Regulations (full title *Regulations for Electrical Installations*), produced by the Institution of Electrical Engineers.

The current edition (first published in 1981) is the fifteenth, which has some significant changes over the previous (fourteenth) edition. As with Building Regulations, changes in the Wiring Regulations are not retrospective and are intended for new installations, but substantial additions or alterations to a house's wiring may mean making changes to an existing installation.

Electrical work should, in general, be left to someone who has the knowledge and confidence to carry it out and who can follow the Wiring Regulations. If that doesn't sound like you, use a qualified electrician who is a member of the Electrical Contractors Association (ECA) or on the roll of NICEIC (National Inspection Council for Electrical Installation Contracting).

The Index at the back of the book will direct you to the details, but if you are only aware of a *symptom*, this checklist may help you to diagnose the plumbing problem.

Basin or sink not clearing
The most likely reason is a blocked waste pipe. Usually, a plunger will clear it (page 74), unless the pipe is frozen (page 40). Also check gullies and drains to make sure they are not blocked.

Cistern slow to fill
If the WC cistern takes a long time to fill, it is possible that a high-pressure type of ballvalve has been fitted instead of a low-pressure type (page 38) or that there is dirt in the valve orifice. Change or clean the ballvalve.

Gutters dripping
Unless the gutter has split or cracked, it is most likely that either the gutter (or downpipe) is blocked and needs clearing (page 78) or that a bracket has given way and the gutter is sagging. Replace the bracket or repair the gutter.

Hot water too hot
Check whether an immersion heater thermostat is defective. If the water temperature doesn't change when the thermostat is adjusted, it should be replaced (page 53).

Overhot water is a characteristic of 'gravity' (i.e. non-pumped) hot water circuits. Consider fitting a thermostatic valve (page 51) or upgrading to fully-pumped operation (see Part Three), which will also improve heat-up time.

Leaky cistern
Many old galvanised cisterns will have rusted sufficiently for the water to be leaking through and showing damp patches on the ceiling. The best thing to do is to replace the cistern completely (page 43).

Living room warm enough, rest of house too cold
If the room thermostat is in the living room and there is additional heating in there (an open fire or gas fire, perhaps), the central heating is turning off too early. Try repositioning the thermostat in another room (page 154).

Noisy cistern filling
This often happens where old-fashioned ballvalves have been fitted. Replace with the correct type of modern ballvalve (page 39). Note that silencer tubes are no longer allowed.

Noises in pipes
Creaks and squeaks in plumbing pipes can be caused by hot water or central heating pipes expanding as they heat up. Find the source of the noise and slip some foam insulation underneath.

Louder banging is likely to be water 'hammer', caused by water being shut off too quickly when a tap or valve is closed. It can usually be prevented by replacing tap washers (page 32) or changing the type of ballvalve used on cisterns (page 38) or closing down stopvalves.

Noise may also be caused by *scale* in direct hot water systems (page 48) or from a faulty boiler (seek professional advice).

Overflow dripping
There are several reasons why an overflow pipe may be dripping or giving a steady stream of water. Check the following:
- [ ] punctured float (page 39)
- [ ] float arm needs bending down
- [ ] float arm catching on cistern
- [ ] faulty ballvalve
- [ ] dirt or grit in the ballvalve.

If the ballvalve is faulty it is likely that either the washer has worn or that the seating needs replacing (page 46).

Pipe joint leaking
Leaks at compression joints (page 23) can usually be cured by tightening the nuts (make sure the pipe hasn't pulled out of the sealing ring); curing a leak in soldered capillary joints and solvent-welded joints in plastic pipes is more difficult and usually involves draining and remaking the joint (pages 22 and 28).

Radiators cold at bottom
This indicates a build-up of sludge due to corrosion. Turn off the heating and drain the system (page 154) before removing the radiator to take it outside for a thorough flushing out with a hosepipe. Refill using corrosion inhibitor (page 155).

Radiators cold at the top
This means either that air has got into the system or that corrosion is causing hydrogen gas to collect. Either can be got rid of by bleeding the radiator at the air vent (page 154), but if this needs doing frequently, find the cause of the air getting in or the corrosion and cure.

▣ Radiator leaking
Most likely caused by corrosion of radiator. Small cracks can be repaired with a leak sealant put into the feed-and-expansion cistern or with epoxy resin filler after draining down, but if damage is serious the radiator will have to be replaced (page 150). Cleanse the system and add corrosion proofer.

▣ Rest of house warm enough, living room too cold
Try *balancing* the radiator first (see page 155). If this doesn't work, there is insufficient radiator area for the living room – perhaps the system was designed assuming there would be an extra heater in there. Try replacing the existing radiator with a larger one (double panel rather than single or one of large area) and perhaps adding double glazing to the windows. An additional radiator may be necessary.

▣ Shower too weak
Unless the holes in the shower rose are blocked, a weak shower is usually caused by insufficient 'head' of water. Cure this by fitting a shower booster pump or raising the cold water cistern (page 107).

▣ Smell from drains
If the drains are blocked they will undoubtedly smell, but there will be other symptoms of the blockage as well. A smell near the property boundary is likely to be a blocked interceptor trap which is allowing dirty water to get to the main sewer but is creating sludge in the bottom of the inspection chamber. Unblock the trap (page 75).

▣ Stopcock won't turn off
Stopcocks tend to get seized up if not turned off occasionally. Apply penetrating oil to the spindle and gentle heat (boiling water) to free them. To prevent it happening again, turn the stopcock on and off at least once a year and close it half a turn from fully open.

▣ Tap cover won't unscrew
Apply gentle heat (boiling water) and a spanner or gripping wrench protected by a cloth to turn the 'self-clean' cover (page 32).

▣ Tap doesn't work
Unless the mains water supply has failed and the cold water cistern emptied, the cause of a tap not working is either a frozen pipe (page 40) or an airlock in the pipe (page 50). With hot taps, it could be scale – see *Water slow*.

▣ Tap dripping
The most usual reason for a tap to be dripping from the spout is a worn tap washer. In extreme cases, the seat may be worn. See pages 31 and 32 for replacing a washer and recutting a seat.

If a tap is dripping from the top when turned on the cause is probably failed packing or a failed O-ring seal. Also page 32.

▣ Tap head won't come off
First check that you have removed any holding screw – they are sometimes hidden under plastic discs; if the head still won't come off, unscrew the cover and place pieces of wood underneath it and close the tap (page 31).

▣ Tap stiff
The most likely cause of a stiff tap is that the packing is too tight – unless the tap spindle is bent. Remove and replace (page 32). Taps should not need force to turn them off – if they do the washer needs replacing or the seat needs recutting/replacing.

▣ Water dirty
Most likely caused by dirt and debris in the cold water cistern. Turn off the water, drain down and bale out the cistern to give it a good clean. Fit an airtight cover or a 'Bye-law 30' kit – see page 45.

▣ Water slow
If the water coming out of taps (mainly hot ones) slows over a period of time, the problem is likely to be scale in the pipes – especially if you have a direct hot water system (see page 48).

Replace badly furred-up pipes (or get a specialist firm in to descale the system) and consider changing to an indirect system and/or installing a water softener (page 42).

▣ WC blocked
Usually the blockage will be in the trap of the WC itself and can be removed by plunging (page 75).

▣ WC won't flush
If a low-level WC cistern refuses to flush, the problem could either be a broken link, a failed washer, or too low a water level. With a high-level cistern, the likely problem is dirt. See page 105.

Dual-flush cisterns (page 103) may have an additional washer which needs to be replaced.

1

PLUMBING BASICS

CHAPTER ONE

THE PLUMBING TOOLKIT

Many of the tools used in plumbing are also used in other types of work around the house. Spanners, for example, are used in car maintenance; hacksaws in metalwork, screwdrivers in electrical work and so on. If you are going to take plumbing seriously, however, it's not a bad idea to keep a separate set of tools in their own tool box – so they are always there for plumbing jobs when you need them and you don't find them covered with oil and grease.

For completeness, the remainder of this chapter is devoted to an A to Z list of plumbing tools, materials and equipment that you are likely to need for the majority of jobs around the house. Some of the more specialist equipment can be hired: see the

notes about hiring towards the end of the chapter.

The ways in which the various tools are used are described here or in the various chapters throughout the book – cutting and bending pipe, for example, in Chapter 2; the correct way to use an immersion heater spanner in Chapter 4.

Remember that, as with tools for any kind of job, there are good tools and not-so-good tools and if you buy poor quality tools, they may let you down at a vital time. This doesn't mean that you have to spend a fortune on tools, but beware of cheap spanners and screwdrivers which may not be strong enough for the job or of cheap hammers which may lose their heads.

Buying the right quality of tool isn't, on its own, enough. The tools have to be looked after which means, in particular, preventing them from rusting (a thin smearing of oil does the trick); tools with cutting edges should be kept sharp with their blades protected against damage.

Never misuse tools: nuts, for example, are meant to be undone and done up with proper spanners. Using pliers or one of the various gripping wrenches will not do the tool much good (they're usually meant for something else) and will probably make a mess of the nut. Equally, gripping wrenches (such as Stillsons) are intended for iron pipes and fittings. If used on copper, they will chew it up.

An A to Z of tools

Adjustable spanners

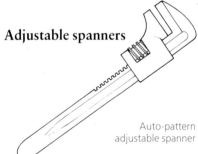

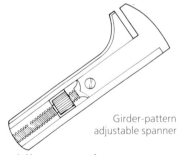

Auto-pattern adjustable spanner

Girder-pattern adjustable spanner

Crescent-pattern adjustable spanner

Many plumbing fittings have nuts on them – compression joints, for example. Unfortunately, there is little standardisation between nut sizes

from different manufacturers – even between fittings for the same size of pipe – so you would need quite a large selection of open-ended span-

ners to cope with all the nuts you might find. For this reason, the easiest way to undo and do up nuts is with adjustable spanners.

There are three main types: auto-pattern, girder-pattern and crescent-pattern. **Auto-pattern** adjustables (used on cars) and **girder-pattern** spanners have slightly different mechanisms, but both suffer from the same disadvantage as far as plumbing is concerned, which is that they are difficult to use on a pipe close to a wall. The better choice here is a **crescent-pattern** spanner which has angled jaws so that it can be reversed in tight corners. Since two spanners are invariably needed for tightening or loosening a joint, the best solution is probably to have two different designs.

Bath/basin spanner

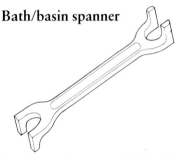

Getting at the nuts which hold the taps in place under baths (or basins) is often awkward. Here a bath/basin spanner (also called a **crowsfoot** spanner) is needed – it comes in two sizes. If the nut is really tight, a metal bar can be used to increase the twisting force on the spanner, but there is a considerable risk of cracking ceramic basins by doing this; try the *gentle* application of heat to loosen the nut first.

Basin wrench

Another type of bath/basin spanner – called a **basin wrench** – has two sizes of serrated jaw. The head is joined to the handle with an adjustable joint so the tool is good for getting into almost inaccessible corners. The good grip and long handle also enable it to move most nuts that other tools won't budge.

Bending spring

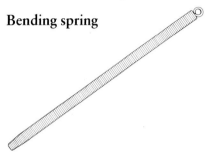

The type of copper pipe most commonly used in plumbing is not easy to bend by hand. Not only is it quite tough, but the pipe will distort as you bend it. A bending spring is slid down inside the pipe for support as it is bent across the knee. Bending springs have a loop on one end (so that they can be pulled out of the pipe afterwards) and come in different sizes for different pipe diameters. They can be hired.

Bending machines

If you have a lot of pipe to bend or if you want to bend stainless steel or 28mm copper pipe (both too tough to bend by hand), the best answer is to hire a bending machine. These machines usually cope with two or three different sizes of pipe.

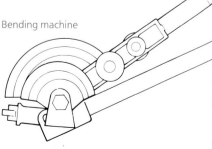

Bending machine

Blowlamp

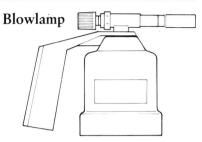

For soldered capillary fittings, a blowlamp is necessary. Modern gas blowlamps are cleaner and safer to use than the old-fashioned paraffin type and, for most jobs, the type which is fitted on to a disposable cartridge is perfectly adequate. An alternative is a **blowtorch** connected by a hose to a cylinder of gas. This is more expensive to buy, but the torch part is lighter to handle than a blowlamp and you don't have to worry so much about running out of gas halfway through a job. Although you have to cart the cylinder around with you, there is a bonus in that a blowtorch can be used upside down to get under a pipe which is being fitted in situ; turning a cartridge-fed blowlamp upside down can cause problems as it will flare.

A blowlamp (or blowtorch) is also useful where a nut seems impossible to dislodge from its fitting, due usually to dried-up jointing compound. Heat will sometimes free it. Don't, however, use a blowlamp for unfreezing pipes or nuts close to ceramic basins or pressed steel or plastic baths: a hairdrier (or variable-heat hot-air gun on its low setting) may be more appropriate.

Hot-air guns can also be used for making soldered capillary joints: special **reflector** nozzles are available for most brands to spread the heat around the pipe.

The electrically-operated Antex Pipemaster is another tool which makes soldering joints safer and easier. But when *Which?* tested it in 1988, the 'safety' stand got very hot in use and had no handle.

Always have a **fire extinguisher** handy when working with a blow-lamp. Failing that, a wet towel can smother an accidental fire. You can get (non-asbestos) mats to protect woodwork.

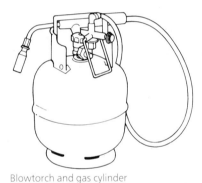

Blowtorch and gas cylinder

Cold chisels

To run pipes through walls can mean making medium or large holes in them. An electric drill fitted with a masonry drill bit can be used to make a start, but sooner or later, you may have to resort to a cold chisel and large 'club' hammer.

Core drill

The ideal tool for making large holes in walls is a core drill, which is rather like a hole saw with tungsten carbide-tipped teeth. Available for hire in a choice of sizes, a core drill will take out a 'plug' of masonry with little damage to the wall. It needs a heavy-duty hammer drill to drive it.

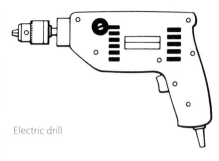

Electric drill

Drills

An **electric drill** is essential for plumbing work: there are a lot of holes that need to be made in walls – either to pass pipes through or to secure pipe clips or to support fittings such as basins, central heating boilers and so on – and making these is much easier with an electric drill fitted with a masonry drill bit. Choose one with two speeds (or variable speed) and hammer action.

A **hand drill** has its uses, too: making holes in tanks, for example, is sometimes a job best done at low speed – see *hole cutters*.

Files

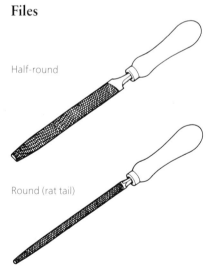

Half-round

Round (rat tail)

Files are used in plumbing principally for getting rid of burs on the ends of cut pipes and for making the pipe ends square. The two most useful types are a small round **rat tail** file and a second-cut **half-round** file.

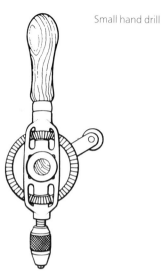

Small hand drill

Floorboard lifter

When you need to get at the space under floorboards, some kind of tool for lifting the floorboard is essential. Good choices include a **bolster chisel**, a car tyre lever or a broad jemmy. If the tool is too narrow, it will damage the floorboard.

Floorboard saw

When a floorboard is difficult to lift or when you only want to take up a short piece, a floorboard saw is helpful. It might be necessary to lever up the floorboard while it is being cut and if the floorboards are tongue-and-grooved, some tongues need to be cut off. The best tool for this is a circular saw set so that the blade only just protrudes beneath the floorboards – beware of electric cables below!

Flux

This is needed for making soldered capillary joints. It is essential to use flux approved by the fittings manufacturer. Some fluxes, especially 'self-cleaning' ones, are highly corrosive and even a small amount left inside the pipe can cause corrosion problems – particularly in central heating systems.

Footprints

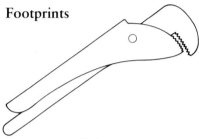

Sometimes called 'pipe tongs', this tool (named after its original manufacturer) is one of the army of gripping *wrenches* used for gripping and turning round objects such as iron pipes (but not copper).

Freezing kits

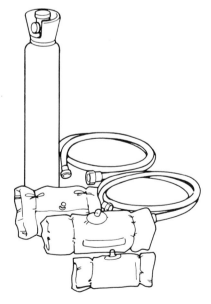

When you are doing small jobs, such as inserting a tee junction into a pipe, you may be able to do this without draining the whole system down. A pipe-freezing kit makes two plugs of ice either side of the joint so that you can cut through the pipes. They can be hired: make sure you have enough freezing 'gas'.

The Drayton Drain Easy kit consists of a stopper which fits into the cold feed pipe outlet of a cistern and a plug which fits over the end of the open vent pipe to allow, say, a radiator to be removed from a central heating system without draining down the whole central heating system, which could well mean losing all your corrosion inhibitor.

Hexagon keys

A range of hexagon (Allen) keys is useful: small ones for jobs like removing grub screws on mixer taps; large ones for installing radiators.

Hole cutters

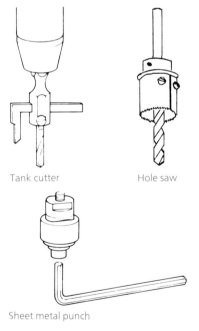

Tank cutter   Hole saw

Sheet metal punch

When making connections to water cisterns for taking pipes or ballvalves, the size of hole you need is somewhat more than can be coped with using a conventional drill bit. There are various types of hole cutter for making medium to large holes.

A **tank cutter** has an adjustable arm which scribes the hole and is best used with a carpenter's brace. A **hole saw** looks like a hacksaw blade curled into a circle and fits on to a twist drill (typically 6mm) and can be used with an electric drill. You need a different size of hole saw for each size of hole: you can buy multiple hole saws which have 'nests' of different sizes. A **sheet metal punch** is drawn through by being tightened with an Allen key. It makes a very neat hole, but you need a different punch for each size.

The tedious way to make a large circular hole is to mark the size of the circle, drill lots of tiny holes around it and file the circle smooth. Sometimes, this is the only answer if the hole has to be a particular size not covered by the other cutters you've got.

Immersion heater spanner

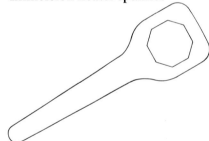

Electric immersion heaters fit into a large screwed boss on the top or side of a hot water cylinder. They are provided with an extremely large nut (58mm or 2¼in), which needs an extremely large spanner to tighten it up – even the largest adjustable won't cope with this. You can hire the correct immersion heater spanner, but their cheapness probably makes buying one a better answer.

Jointing compound

When making fittings where the seal depends on the screwthreads, a jointing compound will be needed. You will find it sold under various brand names. It is used in conjunction with hemp on galvanised iron pipes designed for use with water. Care should be taken to avoid getting jointing compound inside pipes. See also *PTFE tape*.

Measuring tape

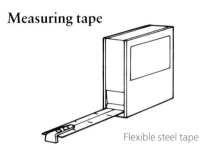

Flexible steel tape

It is helpful to have both a rigid rule – a 1m wooden or steel rule, say – and a flexible **steel tape** for measuring pipe runs and so on. You will have to decide whether you are happier working in metric or imperial units; it's probably best to have a steel rule graduated in both and the longest you can have – 5m or 16ft is a useful length.

Open-ended spanners

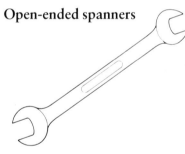

These can be useful for compression joints, provided they are the right size. You will need one spanner for the nut and another one to prevent the fitting turning. *Adjustable spanners* are generally more useful.

Universal plumbing spanners are available with all the sizes you need and including angled basin-nut jaws.

Pipe cutters
A pipe cutter is the quickest tool for cutting copper pipe in the neatest way. Not only does it not create metal filings and a sharp burr on the outside of the pipe (as does a hack-

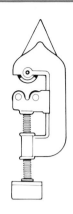

saw), but it also gives a neat square finish to the end of the pipe. Most pipe cutters have a tapered reamer on the end (or in the middle) which must be used after cutting the pipe to remove the slight burr on the inside. All pipes (and fittings) should be smooth on the inside.

For pipes close to walls, look for Pipeslice rotating cutters.

Pliers

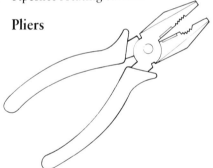

In addition to the various gripping wrenches, a pair of general-purpose pliers is always useful. Choose the electrician's type with rubber handles – they're more comfortable and easier to grip. A pair of **pincers** is useful for removing floorboard nails.

Plunger
If you have serious drain or waste blockage problems, you may need to use specialist equipment or to call in a specialist drain clearing firm. But the common-or-garden sink plunger should be part of any plumber's tool kit – or the household emergency kit.

PTFE tape

Used for wrapping around the threads of screwed fittings, PTFE tape is an alternative to *jointing compound* and hemp. It comes in a roll rather like sticking plaster and is simply wrapped round the threads in a clockwise spiral, following the direction of the thread in several layers.

Radiator key

Also known as a **bleed** key, this small tool is essential for letting air out of radiators. Buy a sturdy brass one – alloy ones can twist and break.

Saws

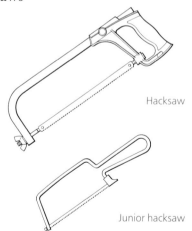

Hacksaw

Junior hacksaw

Although *pipe cutters* are generally more useful for cutting new copper pipe, a **hacksaw** is useful for waste pipe and for cutting through old pipe – especially if it is tough or against a wall. A **junior hacksaw** is useful in confined spaces and generally makes a cleaner cut, though it can be hard work to use.

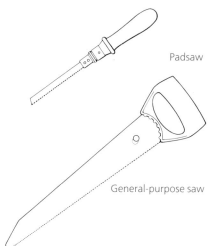

Padsaw

General-purpose saw

The other types of saw which are useful are a **padsaw**, which can be fitted with a hacksaw blade (it cuts on the *pull* stroke) and is useful where space is limited, and a **general-purpose** saw which will cut through metal as well as wood.

Screwdrivers

You probably have a reasonable selection of screwdrivers already. The ones you need for plumbing include a small one for tiny grub screws (on taps, for example), a medium-size one (100mm, say) for screws on WC cisterns and a large one for screwing into walls. You may also need a cross-head screwdriver for some fittings: choose a Phillips type which will fit all cross-head screws.

Self-grip wrench

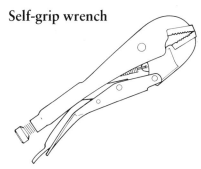

Also known as 'Molegrips', this tool can be locked on to whatever it is gripping, thus giving you an extra pair of hands. The latest design has slightly curved jaws, so can be used more effectively on pipes. It can also be used on stiff nuts.

Solder

You will need a supply of solder for end-feed fittings and it can sometimes help with solder-ring (Yorkshire) fittings as well. You will also need *flux* for making soldered joints. Note that capillary joints use ordinary electrician's solder; plumber's solder is only used for 'wiping' lead joints – a job probably best left to a plumber. Capillary fittings are now made with non-lead (tin/silver) solder: see Chapter 2.

Spanners

The only types of spanner you are **not** likely to need in plumbing are ring spanners and socket spanners. You will need *adjustable spanners*, *open-ended spanners* and gripping *wrenches*.

Stillsons

The largest – and possibly the most useful – of the gripping *wrenches*, Stillsons are designed for gripping round pipes. They should *not* be used for turning nuts which are to be used again. Nor should they be used on copper pipe as they will damage and distort it. But when it comes to removing old steel pipe from a galvanised hot water tank, the largest pair of Stillsons you can get your hands on may be needed. Have two pairs in your toolkit – 8in and 14in. If you need a larger pair (24in, say), hire them for the day.

Tallow

Tallow is used by plumbers as a flux when 'wiping' joints in lead pipe.

Tap reseating tool

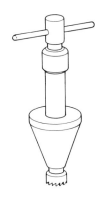

This is a special tool used for recutting washer seats on taps and valves. It can be hired.

Vice

A vice is a useful but not essential piece of equipment. It helps for holding pipes when making joints or for holding fittings which you are dismantling. Vices with pipe jaws are less likely to distort copper pipe: a **pipe vice** can usually be hired.

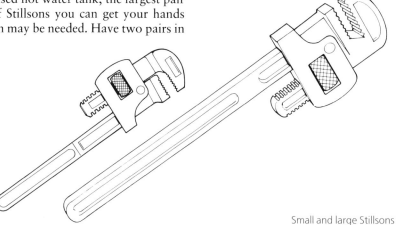

Small and large Stillsons

Waterpump pliers

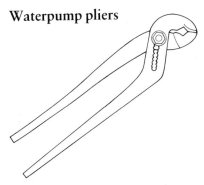

Also known as gland nut or slip-joint pliers, these pliers, unlike other gripping wrenches, *are* useful for turning nuts as the serrations will grip the nut safely. Waterpump pliers can be set to different jaw openings.

Wire brush

Cleanliness is important when making a pipe joint and the first stage with elderly pipes may be to remove rust, old paint and congealed jointing compound. A stiff wire brush is the answer.

Wire wool

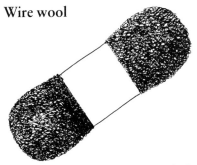

For giving a copper pipe and the inside of the fitting a final clean — vital with a soldered fitting — wire wool is the easiest thing to use. You can, however, get circular **de-burring** tools which achieve the same effect.

Workbench

A flat working surface is always a help. A portable bench incorporating its own vice is particularly useful.

Hiring tools and equipment

The amount of money worth spending on tools for plumbing really depends on how much work you intend to do. It is certainly better to have your own tools wherever possible — at least those necessary to cope with emergencies. Where tools and equipment are too bulky to store or too expensive to buy for occasional use, the answer is to go to one of the many hire shops. You will normally be asked to pay a deposit for tools and equipment (often in the form of a cheque which is returned to you when you return the goods). You will almost certainly be asked to provide some kind of identification when you go to the hire shop — and don't be surprised if you have your photograph taken! Make sure that the tools and equipment are clean and in good working order. If you're not sure how to use something — a pipe-bending machine, say — don't be afraid to ask the hire shop to give you a demonstration. You will have to pay for 'consumable' materials, such as butane gas canisters, carbon dioxide cylinders (for pipe freezing kits) and so on.

You may also have to pay an extra charge if you take equipment back with more damage than when you took it out (make sure any damage is recorded when you pick up the goods) and if the equipment is so dirty that it requires cleaning.

Hiring by the week is usually a little less than twice the cost of hiring by the day. Unless you are certain that you will finish the job in 24 hours, go for hiring by the week. Things can and do go wrong and you don't want to pressurise yourself because you have to get the equipment back to the hire shop. Most hire firms will, however, extend the period of hire, unless the equipment has been promised to someone else.

Buying plumbing goods

There are two main places to buy pipes, fittings and other plumbing supplies. One is a plumbers' merchant which specialises in plumbing goods. Although originally intended for the trade, most of these now sell quite happily to the home handyman. The advantage of plumbers' merchants is that they stock some of the more unusual fittings and will often be able to give you advice. Their disadvantage is that the less exotic goods like fittings are rarely on display so you will need to know what you are after.

The other main place is one of the DIY superstores. These now have an excellent range of pipes and plumbing fittings and are well laid out so that you can see exactly what you need. Many components are sold in kits, so you can, for example, buy all the fittings necessary for installing a basin or a complete waste pack for a modern sink. DIY superstores tend to be a little cheaper for plumbing goods, but you will be lucky to get any advice and you may not find unusual fittings.

Plumbing goods are also sold by builders' merchants and by hardware and do-it-yourself shops. These last two usually have a very limited range and can be quite expensive.

The other type of shop to look out for is one described as a 'bathroom centre' or 'kitchen specialist'. Often these have a good range of plumbing supplies and some are, in fact, plumbers' merchants in disguise.

Another method of buying plumbing goods is by mail order — from suppliers of central heating equipment, for example. These firms advertise in do-it-yourself magazines and *Exchange and Mart*.

Many plumbing fittings are cheaper if bought in quantity: 10, 20 or 50 capillary elbows, for example.

CHAPTER TWO

PIPES AND FITTINGS

In domestic plumbing, two different types of pipe are used: one to carry the hot and cold water to the fitting; the other to carry the dirty water away to the soil and waste pipes. This chapter concentrates on the water supply: for more about wastes and drains, see Chapter 5.

In this country, **lead** pipe was traditionally used for the water system (the word 'plumbing' derives from the Latin word for lead), but this has now largely been replaced by **copper**, and an increasing number of houses have a copper plumbing system. Some houses, particularly those built in the 1930s and before, still have a predominantly lead system. The other main type of water pipe found is galvanised iron 'barrel' piping which is recognisable by its screwed ends. This type of pipe has also been widely used ungalvanised in gas plumbing, but modern installations use copper.

Another type of pipe found in some houses is **stainless steel**. This was used when the price of copper went up. It has the advantage that it does not cause corrosion to galvanised cisterns, but is now more expensive than copper and harder to bend and to join. Capillary fittings need special flux.

Recently, **plastic** pipe has become available for home plumbing for both hot and cold water pipes.

Copper pipe

The great advantages of copper pipe over the old-fashioned lead piping are that it is light in weight, relatively cheap and easy to work with. This has meant that domestic plumbing has been brought within the reach of the competent handyman. One disadvantage of copper pipe is that it can cause the loss of the zinc coating from galvanised pipes and water cisterns.

Copper pipe is sold in lengths, typically of 2m or 3m, though you can buy shorter or longer (up to 6m) lengths from plumbers' merchants. Three common sizes are used for domestic plumbing – 15mm, 22mm and 28mm, all known as 'smallbore'. Plumbers refer to these as 'fifteen mil', 'twenty-two mil' and 'twenty-

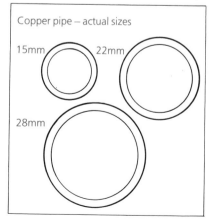

Copper pipe – actual sizes

15mm   22mm

28mm

eight mil', but are equally likely to call them by the names of the pre-metrication pipes they replaced: ½in, ¾in and 1in.

Mathematicians among you will notice that these are, in fact, not exact equivalents. When copper pipe was metricated, the method of measuring it changed. Under the old

Old ½in and new 15mm pipe compared

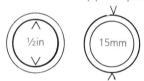

system the *inside* diameter was measured; metric pipe is measured by its *outside* diameter. So old ½in (exact metric equivalent 12.7mm) is in fact the same size as new 15mm, the difference being the thickness of the copper tubing. The same is not so true of 22mm to ¾in and 28mm to 1in where the pipe size has changed slightly as well and you may need to use special types of adaptor fitting.

pipe size	compression fitting	capillary fitting
½in – 15mm	OK	OK
¾in – 22mm	adaptor needed	adaptor needed
1in – 28mm	OK	adaptor needed

Cutting copper pipe

Smallbore copper pipe can be cut with a hacksaw or with a pipe cutter.

A hacksaw (or junior hacksaw) needs to be fitted with a fine blade (coarse teeth will get jammed) and great care must be taken to keep the cut square. A line around the pipe will help: use a piece of paper to draw it. A burr is left on the outside of the pipe: this needs to be removed with a flat file, while a round file is used to clean up the inside. The burr will prevent a proper fitting being made and can also cut your fingers. A hacksaw may be the only possible tool to cut existing pipe which is installed against a wall; however, the disadvantages of a hacksaw are that it produces fine copper filings (and care must be taken not to get these into the pipes) and that it will tend to flatten the pipe slightly – particularly if this is held in a vice. See page 25.

The hardened wheel of a pipe cutter will give a much neater cut: the adjusting handle is progressively tightened as the cutter is rotated around the pipe. The cut should also be square, but a burr will be left on the inside which must be removed with the reamer attached to the pipe cutter. A Pipeslice cutter is designed for use where a pipe is close to a wall.

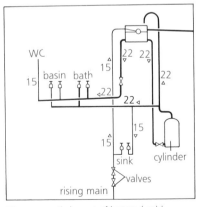

Diagrammatic layout of hot and cold water pipes in typical house showing where the different sizes of pipe are used

Bending copper pipe

Bending a pipe may be necessary when you want the pipe to go round another fitting or to fit through an oblique hole in a wall or ceiling. For turning right-angled bends, a better result is usually obtained using an elbow or slow bend (see page 24). Where you want to bend a short length of pipe, it may be better to use a length of hand-bendable **flexible pipe**. Flexible copper pipes are supplied with plain ends, with solder ring fittings or with tap connectors.

Copper pipe should not be bent without support to prevent it distorting. A **bending spring** can be used for bending 15mm and, if you are feeling strong, 22mm copper pipe. The spring should be cleaned before inserting it into the pipe and a long piece of string attached to the loop on the bending spring. The spring can then be pushed or pulled to where you want to bend the pipe and then pulled out afterwards. The way to bend the pipe is to pull it across your knee, taking care not to 'crinkle' its inner surface as this will trap the spring; protect your knee with some cloth wrapped around the pipe. Getting the spring out is easier if the pipe is slightly over-bent and then pulled back. If part of the spring is out of the end of the pipe, twisting it (with a screwdriver through the loop) can help remove it. See page 25.

Important: where the bend in the pipe needs to be near the end, bend the pipe before cutting it.

If you have a lot of copper pipe to bend, hire a **bending machine**. This tool will also be necessary for bending 28mm pipe. The pipe is placed in the machine with a *former* fitted over it and the handle is then rotated to make the bend. The job is made easier if the bending machine is mounted on its own stand or in a vice and if some light oil is used on the formers. Bending machines are quick and easy to use and can give more accurate bends than springs.

Joining copper pipe

Two main types of joint are used for joining lengths of copper pipe together: the capillary joint and the compression joint. Both are available in a wide range of different forms including straight pipe couplers, elbows, tees, reducers, tank connectors and tap connectors (see page 24 for fittings, page 26 for joints).

Plastic and brass push-fit joints can also be used on copper pipe.

Capillary joints These get their name from the fact that solder is made to flow along the tiny gap between the copper fitting and the pipe by capillary action. Two types are available: *end-feed*, where you add your own solder and *solder-ring* (or 'Yorkshire'), where it comes as part of the fitting. End-feed fittings are cheaper, but solder-ring fittings are easier to use and more widely available. Even when using solder-ring fittings it is as well to have some spare solder

Capillary fittings are now sold with solder based on copper or silver rather than lead. These meet the requirements of the new water bye-laws for pipes supplying water used for drinking or cooking.

The two big advantages of capillary fittings are that they are considerably cheaper than compression fittings and much neater in appearance. You would probably want to use them where the fitting will show.

Once made, capillary fittings cannot be adjusted; this can cause problems if you are putting in a number of elbows.

To make a solder-ring capillary fitting, the ends of the pipe must first be cut square, deburred and thoroughly cleaned with wire wool. The inside of the fitting should be brushed out with a special wire brush and rubbed with wire wool. Don't touch with your fingers after it is cleaned. Then smear a *thin* film of flux over the pipe and the inside of the capillary fitting (very soon after

End-feed capillary fitting

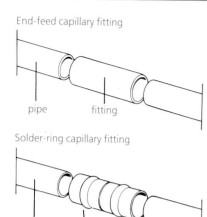

pipe   fitting

Solder-ring capillary fitting

pipe   solder-ring

Compression fitting

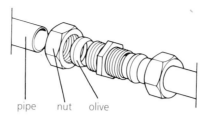

pipe   nut   olive

cleaning) and assemble the two together. Apply flux with your fingers or a rag-covered stick. Flux can be mildly irritating so avoid getting it into cuts or grazes. Then apply heat with a blowlamp, playing the flame gently all round the fitting and adjacent pipe – use a protective mat if working near wooden floorboards which could catch fire. When a bright ring of solder appears at the end of the fitting, the joint is made. If the metal discolours or the flux turns black, you are overheating the joint. Use a mirror if you can't see the back of the fitting. Note that heat will travel along the fitting, so it is usually best to make both ends at the same time – otherwise soldering the second may unsolder the first, unless it is covered with a damp cloth.

When the joint is made, leave it to cool and wipe a damp cloth over the joint to remove any residual flux. If left, this will turn green and will 'bleed' through any paint.

End-feed capillary fittings are made in a similar way: when the fitting has been heated, the blowlamp is removed and solder 'touched' to the exposed end. If the correct temperature has been reached, it will flow into the joint.

If a capillary joint leaks, you *may* be able to feed in more solder (after draining the system). If this doesn't work, replace the joint.

Compression fittings These get their name from the fact that the seal is made by compressing a soft brass or copper ring (called an 'olive') between the fitting and the pipe. Compression fittings are made in both brass and gunmetal.

The main advantage of compression joints is that they make complicated pipe runs much easier since the joints can all be assembled, tightened, slackened off again and the pipe rotated in the joint to get the position right before the nuts are all re-tightened. Although capillary fittings can be 'dry assembled' and the position of pipes marked, they can't be adjusted in this way after installation. Compression fittings can also be dismantled if you want to add to the plumbing system or replace a length of pipe at a later date. It's a good idea to have some spare olives – note that the design varies slightly from one brand to another and that some olives can be fitted only one way round. Note also that you will need a special adaptor compression fitting when joining 22mm pipe to existing ¾in pipe.

To make a compression fitting, the pipe is prepared in the same way as for a capillary fitting, though it doesn't have to be so scrupulously clean. The fitting is assembled by putting the nut over the pipe first, followed by the olive and then the fitting. When making compression joints in vertical pipes, clothes pegs will prevent the nuts and olives sliding down.

To make the joint, the nut is screwed on to the fitting using one

spanner to turn the nut and another to hold the body of the fitting, making sure the pipe is pushed fully home against the 'stop'. This can be difficult in a confined space. It is important that compression fittings aren't tightened too much or the olive will be distorted and the joint will leak. Experience will tell you how hard they have to be tightened: generally 1¼ turns after they are hand-tight is sufficient for 15mm fittings (1 turn for 22mm). A little oil or petroleum jelly on the thread will help. Don't worry if your compression joints leak slightly at first – this happens to professional plumbers too. Just give the nuts a tweak.

Push-fit fittings With brass Spring 'O' fittings, the pipe is simply pushed into the fitting: the seal is made by a rubber 'O' ring and the pipe held in place by a stainless steel retaining ring. The fittings are more expensive than compression fittings, but are unobtrusive and allow the pipe to be rotated in the fitting. If required, the pipe can be removed from the fitting.

Plastic push-fit fittings can also be used for joining copper pipe – see *Joining plastic pipe* for details.

Supporting copper pipe

Although it is possible to install a plumbing system without any supports for the pipes, this is not good practice. The pipes should be supported regularly with plastic pipe clips: for 15mm pipe this means every 1.2m (4ft) for horizontal runs and every 1.8m (6ft) for vertical runs (corresponding figures for 22mm pipe are 1.8m and 2.4m/6ft and 8ft). Pipe clips come in different sizes depending on the size of pipe to be supported and are connected to the wall with a single screw (often round-head) passed through the centre and screwed into a wall plug. Space should be left at the end of pipe runs for slight expansion.

The majority of fittings shown on this page are available in both capillary and compression versions for copper pipe.

● means equivalent fitting available in brass push-fit

■ means equivalent fitting available in plastic push-fit – can be used on copper

□ means equivalent fitting available in plastic solvent-weld – CPVC plastic pipe only

a Straight coupler For joining two pipes of the same size together. *Slip* connectors (usually compression) have no internal stops and can easily be inserted into existing pipe runs ●■□

b Reducer For joining pipes of different diameters – either imperial to metric (adaptor) or, say, 22mm to 15mm. ●■□

c Iron coupler For connecting copper (or plastic) pipe to screwed fittings. Both female iron (illustrated here) and male iron available. ●■□

d Straight tank connector For connecting pipe to water cisterns. Has back nut and available mainly as compression fitting. ■□

e Straight tap connector Also known as 'swivel' connector and used for connecting pipe to screwed 'tail' of tap. Has fibre or rubber washer. ■□

f Drain connector Straight connector fitted with draincock to allow the system to be emptied. Usually compression. A similar fitting, called an *air vent plug* is fitted with an air vent for central heating systems. ■□

g Stop end For blanking off a piece of pipe which is no longer being used. ●■□

h Elbow For turning a 90° corner. 135° elbows (or 'obtuse' bends) are also available, usually capillary. ●■□

Pipe fittings for copper and plastic pipe

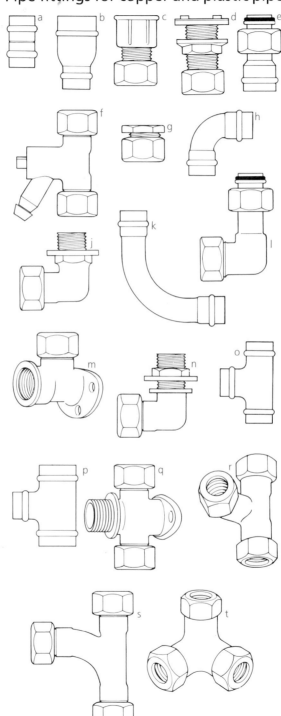

j Elbow, iron one end For connecting copper pipe to screwed pipe or fittings. Both male (illustrated) and female available. ●

k Slow bend For 90° corners where resistance needs to be kept low or to reduce noise. Range of end fittings available.

l Bent tap connector Also known as bent swivel connector and used in same way as straight tap connector (**e**), but allows connection at right angles. ■□

m Wallplate elbow Special connector designed to connect pipework to fittings with male screw threads – such as a bib tap. Has holes to secure to wall. Usually compression. □

n Bent tank connector As straight tank connector (**d**), but allows connection to be made with pipe running at right angles. Usually compression.

o Equal tee Tee connector with all three branches for equal sizes of pipe. ●■□

p Reducing tee Used when one of the branch pipes (to a central heating radiator, for example) doesn't need to be as large as the main pipe. Available with either part of the T reduced. Tee-fittings with *two* branches reduced also available. ■□

q Wallplate tee Similar to wallplate elbow (**m**), this has one iron connection (male or female) and a securing backplate.

r Offset tee Combines a tee fitting with an elbow in one of the branches.

s Sweep tee Used for reducing resistance to water flow. Must be fitted so that sweep is in direction of water flow. Available in different designs – one branch reduced, for example.

t Corner tee For connecting three pipes in, say, the corner of a room. Available with branches equal or reduced.

Cutting copper pipe
Using a hacksaw

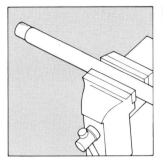

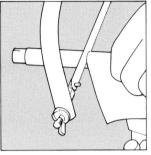

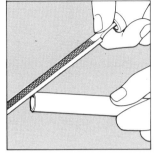

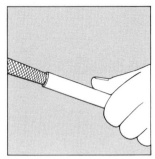

Mounting the pipe in a vice makes it easier to cut, but be careful not to crush it. A piece of dowel inside will support it

Wrap a piece of paper around the pipe to act as a guide and cut through using gentle strokes. Let the saw do the work

Use a flat file (or the flat side of a half-round file) to clean up the burr on the outside of the pipe so that the pipe will enter the fitting

Use a round rat tail or half-round file for the burr on the inside – all pipes should be smooth on the inside. Clean as below

Using pipe cutters

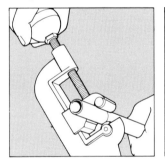

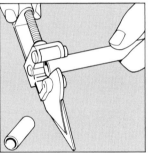

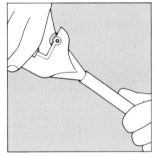

Place the pipe in the pipe cutters with the cutting wheel on your pencil mark. Tighten the wheel down until it touches the pipe

Rotate the cutter once around the pipe and tighten the wheel a little more. Continue like this until you have cut through

Use the reamer on the end of the pipe cutter to remove the burr on the inside of the pipe. This should need only a few turns

Clean up the ends of the pipe to remove grease and dirt. If using capillary fittings, clean the surface with wire wool

Bending copper pipe
Using a bending spring

Flexible pipe

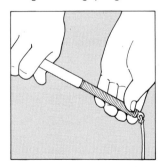

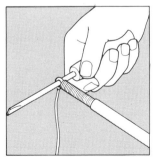

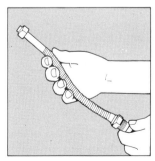

Attach a suitable length of string to the spring and push it into the pipe, twisting it slightly. Position it centrally

Cover the pipe with a piece of cloth and bend it across your knee. Slightly overbend it and bend back to the correct angle

Remove the spring by twisting the end clockwise with a screwdriver. Alternatively, pull it out with the piece of string

This type of pipe can be bent by hand without any trouble. Try to get the correct shape early on – too much flexing will split it

Bending copper pipe
With a bending machine

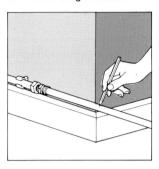

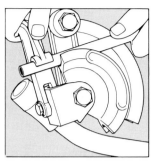

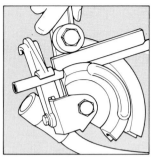

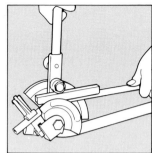

Mark carefully where the pipe is to be bent: do not attempt to make more than one bend at a time without checking

Position the pipe correctly on the curved former of the bending machine, making sure you have selected the right size

Place the straight former over the pipe with the pipe positioned as shown. The concave side of the former should fit the pipe

Move the lever so that the wheel runs along the back of the straight former. When the pipe is bent, open the levers

Joining copper pipe
Making a capillary joint

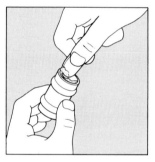

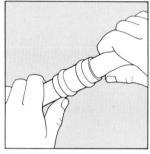

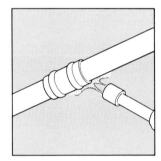

Clean up the ends of the cut pipe thoroughly with wire wool. If the metal is not brightly clean, the joint won't make

Smear flux lightly over the ends of the pipe and the inside of the fitting. Do this with your fingers or a rag-covered stick

Push the pipes into the fitting making sure that they are properly home. Do this fairly soon after applying the flux

Apply heat to the fitting (make sure nothing flammable is around) until a bright ring of solder appears

Making a compression fitting

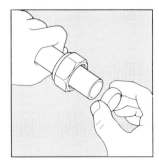

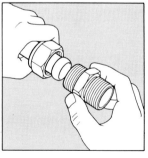

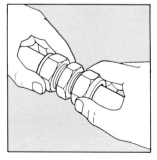

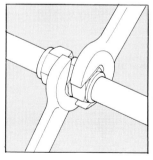

File the ends of the pipe so that they fit, push on the nut and follow with the olive. Smear a little Vaseline on the threads

Push the pipe into the fitting until it meets the internal stop. Prepare the other pipe in the same way and push it in

The nuts should first be done up by hand. Make sure that the pipes stay firmly in the fitting and that the threads are not crossed

Use two spanners to tighten: one on the nut and the other on the fitting. If the joint leaks, tighten a little more

Making a tee fitting

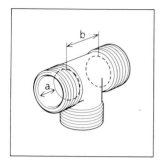

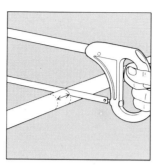

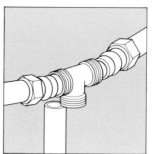

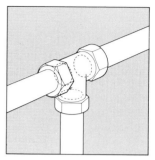

To cut into an existing pipe for a tee, the amount to be removed (b) is the fitting length less twice the insertion distance (a)

Cut out the length 'b', and remove any copper filings left in the pipe. If there's room, use a pipe cutter

Fitting the tee involves some manoeuvring of the two pipes – this may mean removing some of the pipes from their clips

Fix the new pipe and tighten the nuts, being careful not to distort the pipes. A capillary fitting could also be used

Joining plastic pipe
CPVC (Hunter Genova)

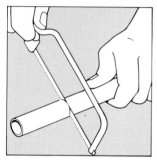

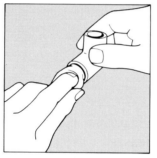

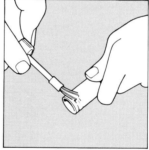

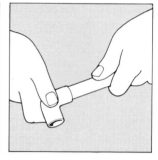

Cut the pipe end square using a fine-toothed saw or a pipe cutter. Remove burrs and rough edges with a knife or file

Check that pipe fits into fitting; clean ends of pipe and inside surfaces of fittings using special solvent cleaner

Brush solvent-weld cement liberally on the pipe and sparingly on the fitting. Work in a well-ventilated space

Push the pipe quickly into the fitting, twisting slightly. Do not remove excess cement. Leave for at least an hour

Polybutylene (Acorn/Polycell)

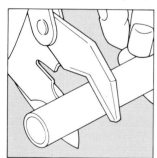

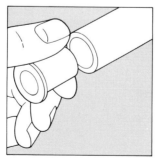

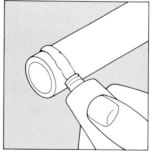

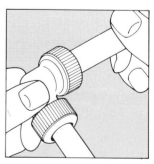

Cut the pipe with a hacksaw or secateur-type cutters; trim rough edges with a sharp knife. (For copper pipe use a pipe cutter)

The inside of polybutylene pipe is supported with a stainless steel insert. Push it in to a depth of 25mm (not needed with copper)

Smear the pipe and the inside of the fitting with special silicone lubricant. Watch out for the internal grab ring – it's sharp!

Push the pipe into place with a slight twisting motion. This requires a degree of force – particularly with copper pipe

Plastic pipe

Although plastic pipe has been widely used for wastes and drains, until recently it has not been suitable for hot water. The reason is that hot water causes the pipes to expand to a greater extent than copper and at higher temperatures it can soften slightly and consequently 'sag' between supports. This is still a problem with modern plastic plumbing pipe, but one that can be overcome if the pipe is well supported and space is left for expansion.

Several types of plastic pipe are available, including rigid cream-coloured CPVC (Hunter Genova), semi-flexible, dark brown polybutylene (Acorn/Polycell) and white cross-linked polyethylene (Speedpex). Plastic pipe can be used for hot and cold supplies and for central heating systems – provided the first length of pipe connected to the boiler is copper. It is available in both 15mm and 22mm (and other sizes) in 3m lengths and can be joined easily to existing copper pipe. Both polybutylene and cross-linked polyethylene are also available in 50m and 100m rolls.

Rigid CPVC pipe is joined by solvent welding. The two semi-flexible pipes can be joined either with plastic push-fit fittings or with brass compression fittings.

Apart from its ease of use, plastic pipe has the advantage of not being so susceptible to frost damage and not causing corrosion with galvanised water pipes and tanks.

Cutting plastic pipe

Plastic pipe can be cut easily with a hacksaw. The burr that is left can be removed with a file or sharp cutting knife. Rigid CPVC pipe can also be cut with a pipe cutter, and polybutylene and cross-linked polyethylene with a special secateur-type pipe cutter.

Bending plastic pipe

Rigid CPVC pipe cannot be bent, though it is somewhat more flexible than copper. Semi-flexible pipe can easily be taken round a 90° corner in a gentle curve: special brackets are available to help with this.

Joining plastic pipe

The wide range of fittings available for rigid CPVC pipe are connected to the pipe by **solvent-welding**. This involves cleaning the ends of the pipe with solvent cleaner and brushing on solvent-weld cement. The pipe is then twisted into the fitting and left for at least an hour while the cement 'melts' the surface of the pipe and the fitting, fusing them together. The solvent-weld cement gives off powerful fumes and should not be used in a confined space. The joint must be made quickly and, if the pipe is a hot water one, be left for four hours.

Polybutylene joints are easier to make as the push-fit fittings are supplied with 'O'-ring seals. All that is needed is a smear of silicone lubricant over the end of the pipe and the inside of the fitting. After a stainless steel insert has been pushed into the end of the pipe, the pipe can then be pushed into the fitting with a slight twisting motion. This requires a little force and might be awkward in tight corners. The rather few fittings available are fairly expensive. Acorn fittings allow the pipe to be rotated in the fitting.

John Guest's white Speedfit push-fit fittings for use with Speedpex need only a clean pipe end, which is simply pushed into the fitting. Disconnection is equally easy, requiring just the grey collar to be pulled back.

Supporting plastic pipe

The intervals at which plastic pipe should be supported depends on the temperature of the water that the pipe will carry, but generally speaking the supports should be positioned at least every 0.5m for horizontal runs and 1m for vertical runs. Space should be left at the ends of the pipe runs for the pipe to expand. Plastic pipe should not be installed near hot appliances such as cookers.

Note that when using plastic pipe the electrical **earth** connection will be lost. The Wiring Regulations require that all metal pipe and exposed metal fittings are connected to the main earth terminal, which necessitates running separate earth wires wherever plastic pipe has been fitted. Where copper pipe has been replaced by plastic or a plastic fitting used to join copper pipe, an earth wire must be fitted across the plastic.

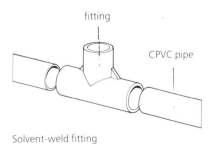

Solvent-weld fitting

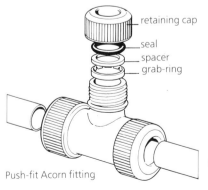

Push-fit Acorn fitting

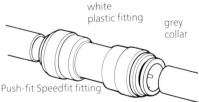

Push-fit Speedfit fitting

Lead pipe

If your house was built before the Second World War it may still have lead piping. Lead piping has two main disadvantages: it is often very small bore, so does not allow a good waterflow, and the amateur will find it difficult to extend the system. Where lead pipe is used for the rising main, it may also contaminate the drinking water – particularly in a soft water area – but there is not much you can do about this since the supply pipe into the house is also likely to be lead and this is difficult and expensive to replace. The installation of lead pipe is now prohibited.

The best thing to do with lead piping is to rip it all out and replace it with copper or plastic. If using copper, replace galvanised water cisterns and tanks; if using plastic, ensure metal fittings in bathrooms and kitchens (sinks, baths and taps) are electrically bonded to earth.

If you do remove all the existing lead plumbing inside the house, you will be left with a screwed connection on the main stopcock to connect to. Old stopcocks were not made to a standard size and it may take some searching in plumbers' merchants to find something that will enable you to connect new copper pipe. There is a fitting (sold for joining washing machine hoses) which has internal and external screwthreads: the external threads are both ¾in BSP (British Standard Pipe); the internal thread is ½in BSP at one end and ⅝in BSP at the other. This may do the trick, but if you are unable to find a fitting of the correct size, you would be best advised to get a plumber in to fit a new stopcock which will involve turning off the water at the water company's stopcock outside your house.

Repairs to damaged lead pipe can be made with a special type of compression fitting – much easier than 'wiping' a lead joint.

Iron barrel pipe

Although galvanised iron pipe is not used in modern plumbing, you may find some connected to old-fashioned hot water tanks. It is possible to buy the screwed connectors to join lengths of this pipe together, but a thread needs to be cut on the pipe itself, which means having special thread-cutting equipment. The threads are known as 'gas' or BSP (British Standard Pipe) threads: those on pipes are referred to as *male* while the internal threads on fittings are known as *female*. Many fittings have tapered threads for a better seal. The traditional way of sealing threaded pipe joints is with *hemp* and jointing compound. This is wound round the threads in a clockwise direction so that it is forced in as the joint is tightened.

Although it is possible to make connections to iron pipe, it is normally considered advisable to replace it with copper or plastic when a new copper hot water cylinder and plastic water cistern are installed.

Hiding pipes

All pipe materials can be painted: copper can be polished up to a fine shine if you like the ship's engine room effect. Alternatively, pipework can be covered up, either with home-made boxing made from plywood or with one of the proprietary products. These include mini-trunking, similar to that used for electric cables, and additional skirting boards and architraves (in plastic or hardwood) which have a space behind them for running pipes. If there is room, fit pipe insulation to both hot and cold water pipes before covering them up.

Where pipes pass through walls and floors, a pipe sleeve should be fitted to protect the pipe and to make the junction look neater. Space should always be allowed for the expansion of hot water pipes.

In solid floors, pipes can be laid in an underfloor duct (**not** embedded in concrete); note that all pipes and joints should be accessible and, where exposed to cold temperatures, well insulated.

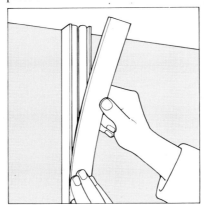
Mini-trunking for covering pipes

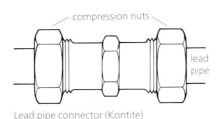
Lead pipe connector (Kontite)

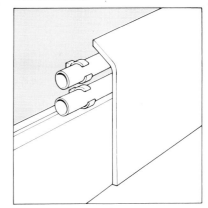
Additional skirting board

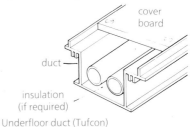
Underfloor duct (Tufcon)

Taps and valves

Taps and valves are used to control the flow of water through pipes, either to turn it off completely or to reduce it. Taps are fitted at the ends of pipes; valves are fitted somewhere in the pipe run.

Types of tap

There are many different designs of tap available and the one you choose will depend on the job you want it to do and the design you like.

Older taps are likely to be the rising spindle kind (see drawing), but these have been widely replaced with shrouded-head taps with non-rising spindles. Many modern taps have ceramic discs, which do not need rewashering and require only ½ or ¼ of a turn to operate.

Single taps Most single hot (red) and cold (blue) taps are *pillar-mounted* with a threaded portion passing through a hole in the sink, bath, basin or bidet and secured by a backnut underneath. The threaded portion is ½in BSP (sink, basin and bidet taps) or ¾in BSP (baths); sink taps are taller to allow room for a bucket underneath.

Bib taps (often used for garden taps) screw directly into a special fitting, such as a wallplate elbow. Many kitchen sinks are fitted with **Supataps**, which can be rewashered without turning off the water. These are no longer made, but replacement washers are still available.

Mixer taps There are three main types of mixer tap: sink mixers, bath mixers and basin/bidet mixers.

Sink mixers have *divided flow* so that the hot and cold water do not mix until they have left the tap. This is important where the cold supply comes directly from the mains (as it usually does in a kitchen) and the hot water comes from the storage cistern. They can be either **one-hole** or

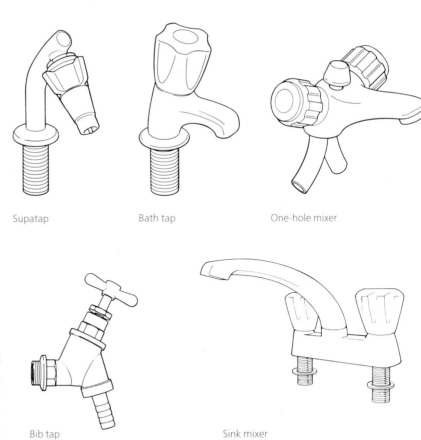

Supatap

Bath tap

One-hole mixer

Bib tap

Sink mixer

two-hole depending on the fitting details for your sink. The standard fitting distance for the inlets on sink mixers is 178mm (7in).

Most bath and basin mixers also have two inlets and one outlet, but the two flows are allowed to mix freely. The difference between a bath mixer and a basin mixer is the spacing of the two inlets: 181mm (7⅛in) apart for baths; 102mm (4in) or 210mm (8¼in) for basins. One-hole mixers are also available for basins and bidets. Bath/shower mixers have a connection for a shower hose and a diverter so that flow can be directed to the shower (see Chapter 8). **Three-hole** mixers have a third connection for a pop-up waste control and/or the spout.

Most mixer taps have screwed BSP connections: some one-hole mixers have plain copper tube – usually 15mm, but sometimes 12mm.

Most mixer taps simply use the hot and cold knobs to control flow and temperature, but some modern one-hole mixers have just a single lever which does both.

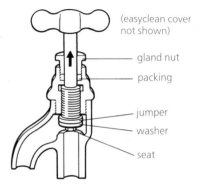

(easyclean cover not shown)

gland nut

packing

jumper

washer

seat

Rising spindle tap with cross-head. As the tap is opened the whole spindle moves

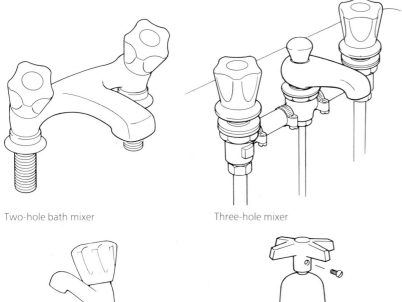

Two-hole bath mixer

Three-hole mixer

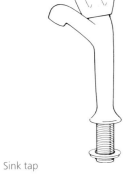

Sink tap

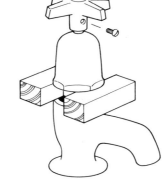

Method of removing tight tap heads

Problems with taps

Taps and valves are essential parts of the plumbing system. It is important that they are kept in good working order so that water does not leak from the system and so that you can turn the water off when you want to.

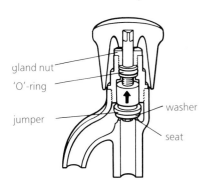

gland nut

'O'-ring

jumper

washer

seat

Non-rising spindle tap with shrouded head. The spindle lifts the washer assembly

The most common thing that goes wrong with taps is that they leak – either because the gland or packing has failed (a leak from the top when the tap is open) or because the washer has worn (a leak from the spout when the tap is closed).

Packing can usually be replaced without the need to turn the water off, but this will require the removal of the tap head and the 'easyclean' cover. Old-type cross heads can be difficult to get off even after the grub screw has been removed. If a gentle tap with a hammer doesn't work, open the tap fully, unscrew the easyclean cover, pack two pieces of wood underneath it and tighten the tap head down. Removing the easyclean cover can be very easy or very difficult. Always protect it with cloth if it needs a spanner or wrench to

shift it. Boiling water or heat can help to shift a stubborn cover.

Traditional packing can be replaced with wool smeared with petroleum jelly; some modern non-rising spindle taps have rubber (or neoprene) 'O'-ring seals which are easily replaced.

When mixer taps leak at the base of the swivel nozzle, the washer or 'O'-ring there needs replacing. The spout usually unscrews or lifts off easily, but special *circlip* pliers may be needed. Lubricate the washer (or 'O'-ring) with silicone grease.

The method of replacing a washer depends on the design of tap. Always keep spare tap washers handy and replace the washers regularly – no tap should need to be turned off with force. Sometimes it may be necessary to replace both the washer and the jumper – particularly if they can't be separated. Except with Supataps, replacing tap washers may mean turning the water off and draining the cistern. Do this by tying up the ball valve (so the kitchen tap still operates) and opening the *cold* taps – even if it is a hot tap which is being rewashered.

Tap washers come in different sizes and different designs (hemispherical ones, for example, are often used on mains pressure taps). If a washer has a brand name on it, make sure that the smooth side comes into contact with the seating.

If the valve seat of a tap has gone, it can be recut or, more simply, a nylon replacement seat can be inserted. Replacement seats cannot be fitted to Supataps.

Tap conversion kits

Many existing taps with cross-head handles can be made to look like new ones with shrouded heads. Some **tap conversion kits** simply replace the existing handle with a new one; others replace the spindle assembly and may well provide a replacement nylon seat.

Rewashering a tap

Rising spindle and shrouded-head taps

With most taps you will have to unscrew the 'easyclean' cover. If you have to use a spanner on this, protect it well with cloth

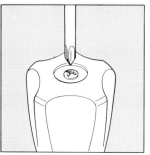

The tops of shrouded head taps can often be removed only by undoing the screw hidden under the coloured disc in the head

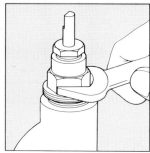

Remove the main nut holding the head gear in place. Use a second spanner (and cloth) to prevent the tap turning

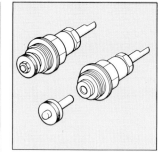

Lift out the jumper (if not fixed) and take off the washer, removing any nuts first. Fit the new washer and reassemble

Supataps

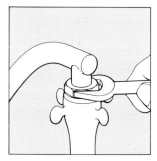

There is no need to turn off the water supply when rewashering a Supatap. Simply unscrew the nut and turn the tap on fully

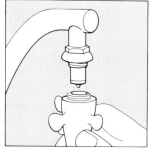

When the tap reaches its fullest extent it will come off (and probably fall in the sink) and the water will stop flowing

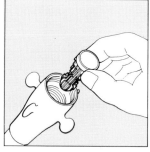

Unless the anti-splash device falls out of its own accord, tap the nozzle on a hard surface to remove it

Use a screwdriver to separate the washer/jumper unit from the anti-splash device and replace it with a new one. Reassemble tap

Repairing a leaking tap

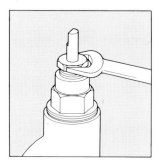

Tightening the gland nut will sometimes stop a leak. If not, unscrew it and remove the packing with a penknife

The tap will be sealed either with Vaseline-covered wool or with a rubber 'O'-ring seal. These should be replaced

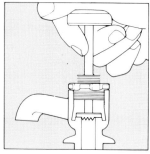

If the valve seat is damaged, it can be recut using a special tool. This fits into the body of the tap and is turned to grind the seat

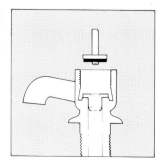

Alternatively, a replacement nylon seat can be fitted which is placed firmly on the old seat and a new washer and jumper fitted

Replacing a tap

How taps are fitted

A basin pillar tap is sealed to the basin surface with non-setting mastic or a supplied gasket and a plastic washer underneath

A sink pillar tap usually has a plastic washer (or rubber gasket) above the sink and a 'top-hat' washer underneath

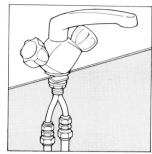

With a mixer tap, each supply has its own connection. Single-hole mixers may have plain pipes for the connections

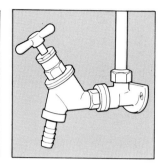

This type of (Bib) tap is easy to connect: it simply screws into a wallplate elbow (use PTFE tape for sealing the threads)

Replacement

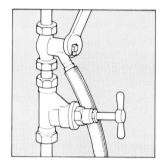

Turn off the water for the section you are working on and open the taps to drain the pipes. In the kitchen, drain at the draincock

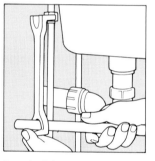

Use a bath/basin spanner to undo the nut on the tap connector. This may need extra leverage, but don't overdo it

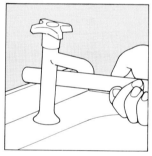

Undo the backnut, with a second person or piece of wedged wood stopping the tap turning. Use *gentle* heat if necessary

Remove the old tap from the basin or sink and scrape away any old dried jointing compound from the surface. Don't scratch it

Fit the new tap, using a plastic washer and 'top-hat' washer if necessary. Tighten the backnut, supporting the tap as before

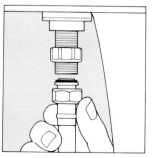

Check the position of the new tap 'tail'. If too short, use a tap adaptor. Fit the tap connector, not forgetting the washer

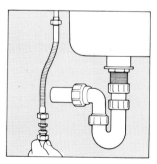

If replacing the old length of pipe, use a flexible corrugated copper tube connector which makes working much easier

Tap conversion kits

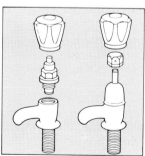

One way of improving the look of old taps is to fit a tap conversion kit in place of the tap spindle or the handle

Types of valve

Valves are fitted in the hot and cold water supplies so that the water can be cut off from the whole system or from individual branches. The more valves you fit in a system, the easier it will be to carry out maintenance or replacement. Never fit valves to safety vent pipes.

There are four main types of valve: stop valves, gate valves, servicing valves and drain valves.

Stopvalves This type of valve is fitted where the water is at mains pressure, so is the type fitted as the main stopcock in the rising main. In operation it is similar to the normal type of tap: the water flows up through an orifice which can be closed by a jumper with a washer screwed down on to it. Stopvalves are stamped with arrows showing which way round they should be fitted and are usually supplied with compression fittings. They have to have water authority approval for use on mains water pipes.

As well as in the rising main, stopvalves are typically fitted in the branch pipe to an electric shower, on the pipe leading to a garden tap, on pipes leading to a washing machine or dishwasher and on the pipes to a water softener. A garden tap may also need a *double-check valve* – see page 122.

Gatevalves Unlike stopvalves, gate-valves only slightly restrict the flow when the valve is opened, so are fitted typically to the cold supply pipes leading from the cold water cistern where the water pressure is lower. They differ in operation: the control mechanism is a type of solid portcullis (or 'gate') which is screwed down across the flow of water. When the gate is raised, the water can flow freely. Gatevalves are usually fitted with wheel handles rather than cross-head handles and must be kept fully open to avoid

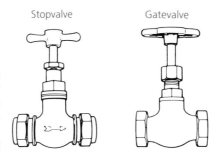

Stopvalve   Gatevalve

airlocks. They can be fitted either way round and are usually supplied with compression fittings.

If the cold supplies from your cold water cistern aren't fitted with gate-valves, this is reasonably easy to do as it requires only cutting out a short length of pipe. To save draining the cistern, make a bung from poly-ethylene sheeting and cloth and hold it against the outlet. Turning on the taps should cause this to seal the opening. Alternatively, use a solid stopper.

Servicing valves It is a requirement of the water bye-laws that servicing valves be fitted on the pipes leading to float-operated valves (ballvalves) – on cold water or WC cisterns, for example. See page 39.

Most servicing valves have an internal ball with a hole through it, so that just a quarter of a turn is needed to move the valve from fully

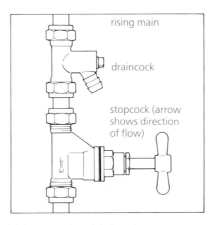

rising main

draincock

stopcock (arrow shows direction of flow)

Mains stopcock and draincock

open to fully closed. Operation is either by a circular knob or with a screwdriver fitted into a slot. Some servicing valves (sold for installing a washing machine) are incorporated into a tee fitting and have a lever to operate.

It is a good idea to fit servicing valves before all taps so that they can be rewashered without the need to drain down the whole pipe (and, possibly, the whole cistern).

Drainvalves Unlike the other three types of valve, a drainvalve (or draincock) is usually kept closed and opened only when the system is to be emptied of water. Drainvalves are generally fitted in the rising main just above the main stopcock and in the cold supply pipe to the hot water cylinder at the point where it enters the cylinder. They are also fitted at the lowest point of a central heating system (and on downloops and, perhaps, next to the boiler) and, usually, just before a garden tap. A drainvalve has a nozzle for a hose pipe to be fitted. Good quality drain-cocks have a gland which prevents water leaking around the top of the fitting when it is open.

Problems with valves

Gland packing can need replacing on valves (same procedure as for taps), but washers rarely need replacing. But the main problem with valves, especially stopvalves, is that they stick open. It is a good idea to open and close valves every so often to make sure that they don't get gummed up. If they do jam, keep applying penetrating oil until they are free. Do not apply undue force: this may damage the valve.

After freeing a stopvalve, open it fully and then close it $\frac{1}{4}$ to $\frac{1}{2}$ a turn. This won't significantly affect the water flow, but should prevent it sticking in the future. This is not necessary with gatevalves or ball-type servicing valves.

CHAPTER THREE

THE COLD WATER SYSTEM

The supply of water in England and Wales is the responsibility of private companies appointed by the government under the Water Act 1989. Ten of these are the successors to the regional water authorities and they also provide sewerage services. The remaining 29 companies, which were in existence before privatisation of the water authorities, serve particular localities and are responsible for water supply only. In Scotland, water is supplied by 12 water councils.

Water is normally paid for via the water rates, but the water industry is looking at the feasibility of water metering in the future.

Cold water supply

The water arrives at your house via the water mains which runs down your road. Each house will have its own branch **service pipe** which will be fitted with a stopvalve, referred to as the **water company stopcock** – usually just outside the boundary of the property. This valve is generally situated under a small metal cover – in towns, it is possible to see a whole row of these stretching down the pavement; in rural areas, it may be more difficult to find. This stopcock may need a special key to turn it on and off. Normally, unless you have one of these keys, you would have to call the water company's engineers if you wanted the mains turned off for any reason – perhaps to work on the mains stopcock inside the house.

From the water company stopcock, the service pipe runs underground (buried at least 750mm down for protection against frost) and enters the house. It is usually protected from damage by being run inside a drain pipe, and rises slightly to avoid air bubbles. After entering the house the service pipe usually terminates in the householder's **main stopcock**, which is where the house supply starts. It is worth noting that the householder is responsible for the service pipe as well as for the domestic distribution system. Service pipes can run the length of the house and may need insulating against frost damage where exposed.

At one time, the water mains and service pipes were made from galvanised iron or lead and were also used for providing an electrical earth connection to the house. Lead and iron were superseded first by copper, which is more resistant to corrosion, and, more recently, by plastic. Because of the increasing use of plastic (which does not conduct electricity), water mains are no longer used to supply an earth connection and Electricity Boards supply a separate earthing terminal within the house for the consumer to use.

The mains stopcock

In many houses, the mains stopcock is situated under the kitchen sink. The reason for this is simple: at least one tap must always be supplied directly from the mains for drinking purposes and this is usually the cold tap in the kitchen. Sometimes, there is a separate 'drinking water' tap. The stopcock isn't always here: in older properties, it might be underneath a floorboard just inside the front door, in a cupboard, in the garage, or there may be no householder's stopcock at all.

It is important that you know *where* the stopcock is and *how* to turn it off. In emergencies, such as a burst pipe or major leak, turning off the stopcock should be the first thing you do even if it has no immediate effect. Stopcocks, like all valves, are turned off (closed) by turning the handle *clockwise*; they are opened by turning them *anticlockwise*. It is a good idea to turn the main stopcock off and on once or twice each year to keep its mechanism free and not wait for an emergency to discover that it has jammed open. The drawing on the opposite page shows a mains stopcock with a draincock fitted above.

Cold water plumbing

From the householder's main stop-cock, the water is taken by the **rising main**. In most houses, with what is called an *indirect* cold water supply, the rising main goes directly to a cold water cistern, typically situated in the loft; in others, with a *direct* cold water supply, the supply for taps and fittings within the house is taken from the rising main itself.

Indirect systems

In an indirect system, the only connection to the rising main usually permitted is the obligatory one going to a drinking water tap in the kitchen (though some existing houses may have other connections – for a WC, for example). Since the drinking supply is often the only cold water supply in the kitchen, you are, however, allowed to take other connections off this branch – for supplying an outside tap or for plumbing in a washing machine. But note that you must use fittings and valves which comply with the water byelaws and that if the cold supply is connected up to a washing machine from the rising main, this will be at mains pressure while the hot supply (if there is one) will be at a lower pressure. See Chapter 7 for details.

The other connection that you might want to make to the rising main is one for an instantaneous gas water heater or electric shower, which need mains pressure, or one of the devices which alter the quality of the water – a softener, conditioner or filter, say.

The rising main passes up inside the house all the way to the large water tank in the loft. It is more correct to refer to this as a cistern, since a 'tank' is, strictly, closed on all sides and under pressure whilst a cistern is open at the top.

The cold water cistern acts as a store of water and feeds the remainder of the house.

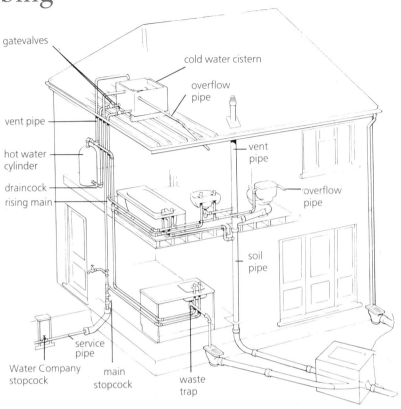

Indirect cold water supply – all cold taps, except kitchen cold tap and garden tap, are fed from the cistern, as is the feed to the hot water cylinder

The cold water cistern

The rising main is connected to the cold water cistern by means of a ballvalve: a valve which operates by the action of a metal (or plastic) ball floating on the surface of the water which will close the valve when the level of water is sufficient, and open it when the level falls. Cold water cisterns are also fitted with an overflow pipe (more correctly called a 'warning' pipe) which sticks out of the house through the eaves; the tell-tale drips or steady stream will tell you when something is wrong.

Note that this ballvalve (unlike other ballvalves in the house) must be a high-pressure one since it is connected to the mains.

The cold water cistern should be fitted with a lid to keep out light and insects but should not be airtight.

As the cistern supplies the remainder of the house, there will usually be two main feeds: one for the cold taps (not the kitchen) and WCs and one for the hot water system.

These feeds can usually be seen as separate pipes leading from the cistern; sometimes there may be extra feeds fitted for a shower or bidet. It is important that all feed pipes from the cold water cistern are fitted with valves so that each of the different water circuits can be isolated and worked on without the need to drain the entire system. Cold water feeds are usually fitted with gatevalves near to the cistern; the one for the cold feed to the hot water

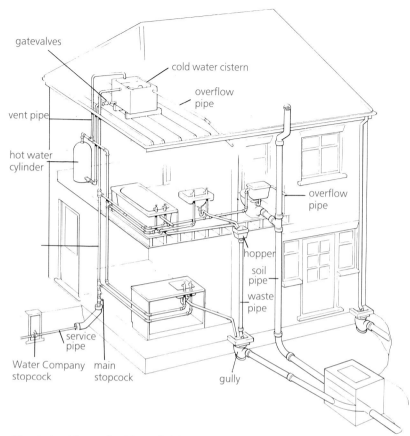

gatevalves

cold water cistern

overflow pipe

vent pipe

hot water cylinder

overflow pipe

hopper

soil pipe

waste pipe

service pipe

Water Company stopcock

main stopcock

gully

Direct plumbing – all cold taps fed from rising main. In this example, a cistern feeds the hot water cylinder

Which system is best?

Although direct plumbing systems are less complicated and cheaper to install, most water companies have preferred indirect systems and many have allowed only this sort.

From the water companies' point of view, the advantage of indirect systems is that they reduce the effects on the mains of heavy demand (first thing in the morning, for example) and that they keep most of the householder's water system separate from the mains supply. This makes it very difficult for the mains to be contaminated by dirty water (from a bath, say) being sucked back up the mains – a phenomenon known as *back siphonage*. Water companies are very concerned about maintaining the quality of their water and many of the bye-laws deal with the prevention of back-siphonage. This is one of the reasons why only approved fittings may be used.

From the householder's point of view, indirect systems have advantages. The cold water cistern provides a reserve of water in case the mains should be cut off for any reason, and because the system (apart from the rising main) operates at a lower pressure it is quieter.

But indirect systems have drawbacks, too – notably that there may be insufficient 'head' to provide a decent shower and that the cistern and pipework in the loft space are prone to freezing. Both these can be overcome with an *unvented* hot water system – see page 57.

system is generally fitted near to the hot water cylinder. To ensure an adequate flow of water, feed pipes should generally be 22mm.

The reason for having separate feeds is one of safety. If a shower (other than the instantaneous type fitted to the rising main) is connected to a tap or shower fitting fed from a single pipe leading from the cold water cistern, flushing the WC or turning on another cold water tap can starve the shower of cold water, making it uncomfortably hot.

Cold water cisterns need to be quite large to cope with the demands of an average house in the event of a mains failure – the normal recommended size is 230 litres (50 gallons) *actual* capacity, which is around 320

litres (70 gallons) *nominal* capacity (filled to the brim). If an existing cistern is too small, a second cistern can be installed alongside joined by two 28mm pipes at low level.

Direct systems

In a direct system, all the cold water pipes and WCs are connected directly to the rising main. There may, however, be a feed-and-expansion cistern if central heating has been installed and, sometimes, a cistern for the hot water cylinder.

Different methods for supplying the hot water may be needed with a direct plumbing system: these include gas 'multi-point' heaters and combination boilers. For more details, see Chapter 4.

Head

The flow of water from any fitting, such as a tap, is related to the water pressure in the pipe leading to it. This is measured as the 'head' or the vertical distance from the fitting to the level of water in the cistern. Because some of this pressure is lost by the water having to pass along pipes, round corners and through fittings, the normal convention is to measure the head to the *bottom* of the cistern.

Water metering

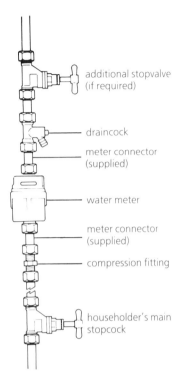

additional stopvalve
(if required)

draincock

meter connector
(supplied)

water meter

meter connector
(supplied)

compression fitting

householder's main
stopcock

Most water companies offer the choice of having a water meter installed so that you pay for the water you actually use rather than pay water rates based on the rateable value of your house. The companies publish information leaflets which give a rough guide to whether you are likely to benefit from having a meter and will tell you whether one is possible for your house.

You have to pay an installation charge for the meter and then a quarterly standing charge plus an amount for each cubic metre (1,000 litres/220 gallons) used.

The chances are that you will save money with a water meter – especially if you live in a house with a high rateable value – and should repay the costs of installation within two years. This period can be reduced if you are allowed to install the water meter yourself. It is fitted just above the main stopcock.

Cisterns and ballvalves

In most installations, the cold water storage cistern and its ballvalve are the heart of the cold water supply. There are other cisterns in the house – the feed-and-expansion cistern for the central heating system and individual ones for WCs, for example.

Cisterns

There are three main types of cold water cistern: galvanised iron, rigid plastic and flexible plastic.

Galvanised iron cisterns are the traditional type. They are extremely heavy and expensive and eventually suffer from corrosion – particularly when used with copper pipe. If old galvanised cisterns are badly corroded, they should be replaced with a cistern made from one of the other materials.

Rigid plastic cisterns (polypropylene) have the advantage that they are the same size as the galvanised cisterns they replace. Their main disadvantages are that they are affected by jointing compound and that they may not fit through the loft hatch. A typical 230 litre cistern, for example, measures 1070mm × 660mm × 585mm though a wide variety of sizes is available.

Although it's less convenient, you could put in two smaller cisterns linked together with two 28mm pipes (fitted 50mm up from the base) rather than one large one. If your present cistern is too small, a second one can be added. The connections should be arranged so that the ballvalve is connected to one cistern and the cold feeds to the other so that the water doesn't become stagnant.

Flexible plastic cisterns are circular and are particularly useful – they can be squashed enough to be squeezed through the loft hatch. On the other hand, they provide no support for the pipework which can cause the cistern to split or make the plumbing noisy if the pipes are not adequately supported in some other way. Jointing compound must not be used with these either.

All new and replacement cisterns have to meet Water Bye-law 30, which has specific requirements for insulation, ventilation, a tight-fitting lid and the various connections. To comply, cisterns are sold with Byelaw 30 'kits' – see page 45 for details.

Ballvalves

A ballvalve controls the flow of water into a cistern. The mechanism is operated by a metal or plastic ball or float which floats on the surface of the water: as the water level rises, the valve is closed, shutting off the supply.

It is important that any ballvalve is the appropriate design for the water pressure. For mains pressure (for example, the cold water cistern connected to the rising main), use a high-pressure ballvalve; for other cisterns which are fed from the cistern (such as a WC), use a low-pressure design. Check whether WCs are connected to the rising main before fitting a new ballvalve.

There are three main types of ballvalve: piston, diaphragm and equilibrium (not widely available).

Piston ballvalves are the simplest. The *Croydon* design used a vertical falling piston against a nozzle and is now obsolete. The *Portsmouth* design has been and still is commonly used in houses. Here, the float arm closes a horizontal piston, fitted with a washer against the valve seating. It is noisy in operation, so *silencer* tubes were fitted to many Portsmouth valves to reduce noise – these fit into the inlet and discharge water below the water level in the

Ballvalves

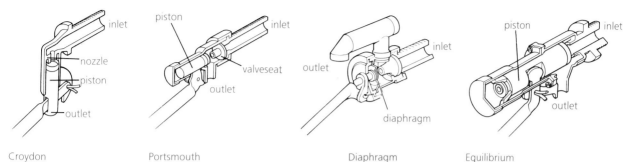

Croydon   Portsmouth   Diaphragm   Equilibrium

cistern. The installation of silencer tubes is no longer allowed.

Diaphragm ballvalves were developed to overcome the noise of ordinary piston types. With this type of valve (of which the *Garston* or *BRS* valve is the best known), there is no moving piston – rather a rubber diaphragm which closes over a nylon seating with the last movement of the float arm. This type is available in either brass or plastic, and often sold as 'quiet' ballvalves.

Equilibrium ballvalves were developed to overcome problems of *water hammer* caused by ballvalves shutting off too quickly. In this type, the flow of water is actuated only by the movement of the float arm – the valve cannot be forced open by the pressure of water in the mains pipe. Once open, the valve is balanced by equal water pressure on the two sides of the piston.

All domestic ballvalves are fitted with screwed inlets (usually ½in BSP) to take a tap connector. Unless used on 'Byelaw 30' cisterns and fitted with a downturn in the operating lever (so the float can easily be adjusted), piston-type ballvalves are no longer allowed on cold water cisterns unless they are fitted with a double-check valve.

All new ballvalves should be provided with a *servicing* valve on the inlet pipe.

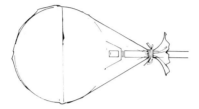

Pinholes in ballvalves can be temporarily repaired with a plastic bag

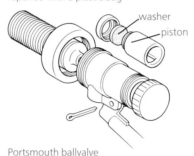

Portsmouth ballvalve

Diaphragm ballvalve

Faults with cisterns

Sediment at the bottom should be cleaned out regularly – after draining the cistern. Galvanised iron cisterns suffer from corrosion, and when this is serious they should be replaced with a plastic cistern.

Faults with ballvalves

The two main things that go wrong are that the ball corrodes or that the valve itself starts leaking.

A corroded ball can 'pinhole' and start filling with water, which will increase the level of water in the cistern. A temporary solution is to enlarge the pinhole, empty out the water and tie a plastic bag over the ball. It should however be replaced as soon as possible.

Leaks can happen because dirt or grit has got into the valve or because the valve washer is worn. This will mean dismantling the valve and extracting the washer (see page 46). Be careful when doing this not to damage the piston. Before re-assembling the valve give it a good clean and smear petroleum jelly on the piston and on the threads of the cap holding the washer, so that if necessary it can be undone more easily in the future.

It is worth checking the operation of all ballvalves regularly, particularly the one in the feed-and-expansion cistern. Because this is not used very often it may be stuck in the closed position over an empty cistern if the contents have evaporated.

If the level of water in a cistern is too high or too low, it can be adjusted by bending the float arm (if it is metal). Take care when doing this not to put any strain on the valve itself. Modern types of ballvalve come with a much easier method of adjustment.

More about cold water plumbing

Pipe insulation

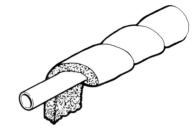

Frost protection

All cold water pipes, from the point where the service pipe enters the house, should be lagged with pipe insulation. The most important ones are often the most inaccessible: under the floorboards near to airbricks and up in the loft space near to the eaves. Modern pre-formed pipe lagging in a range of sizes is generally more efficient and easier to apply than previous types and is relatively inexpensive. Felt or glass-fibre wrap is still probably the best type of insulation for valves and compression fittings where pre-formed insulation won't fit. Cold water cisterns themselves should also be lagged (see page 44): put insulation on the top of the cistern but not underneath, so that some heat comes up from the house to keep the chill off the cistern. When insulating the loft floor, bring the ends of the blanket up against the sides of the cistern to give continuity.

Where a pipe is very exposed – in an outside WC, for example – a small heater will keep the chill off the pipes. Electric frost protection tapes incorporating a low wattage heater, could be wrapped round pipes. They need no special wiring and may be thermostatically controlled to come on at 3°C. An outside cistern could be kept from freezing by hanging a light bulb near it and leaving it on all the time.

If a pipe has frozen, it can be thawed out with a hairdrier or by wrapping towels soaked in hot water around it. You will know which pipe is frozen because the tap at the end of it will no longer be working. Somewhere, there will be a plug of ice stopping the flow. Open the tap and work backwards along the pipe, warming it up until the water flows again. Lag the pipe to prevent it happening again.

If you do have the misfortune to have a burst pipe, the most important thing is not to panic. Always turn off the water at the mains stopcock first (or tie up the ballvalve) and turn on all taps (provided the sink and basin wastes aren't frozen up!) and then attempt to trace the position of the burst. Use a torch to find your way around – the electrical fittings may be unsafe if water has been pouring through them. Often, with copper pipe, frost will have 'blown' a compression joint, forcing the pipe out of the fitting. This is easily remedied, but a new olive may be required. More of a problem is where water has frozen inside a pipe and split the pipe. The answer here is to cut out the offending piece of pipe and to insert a new piece. There are several special fittings for doing this, including a short length of pipe and push-fit brass or plastic fittings. As a temporary measure (especially useful with burst lead pipe), there are resin filler products which can be applied to the pipe to stem the flow of water, some of which will withstand mains pressure almost immediately. Some need to be used in conjunction with tape or bandage. A temporary repair can also be made very quickly with a burst pipe clamp.

Wrap-round

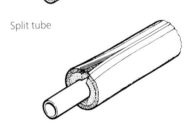

Split tube

With 'zip'

Repairing burst pipe

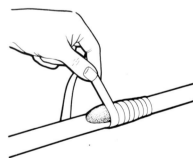

Using tape and resin filler for repairing bursts

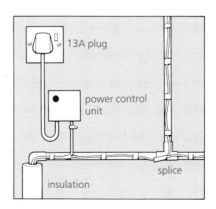

Low wattage frost protection

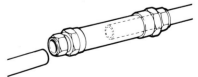

Slip fitting for replacing length of pipe

Electrical bonding

Although mains water pipes are no longer used as the source of a house's electrical earthing, it is still necessary to make sure that all metal pipes inside the house are connected to the earthing point provided by electricity boards as part of their supply.

There are three main points where an earth connection has to be made. First the main water service pipe (and, incidentally, the main gas service pipe) has to be earthed (with $6mm^2$ or $10mm^2$ cable) as soon as possible after it enters the house. Second, an earthing 'bridge' must be applied across plastic pipe or fittings inserted into water pipes anywhere in the house. Third, all accessible metal pipework must be earthed wherever it is within reach of a sink, basin, bath or other water source. This means running an earth wire ($4mm^2$, or $2.5mm^2$ if run in conduit) from the main earthing point and linking it to the hot and cold pipes, the taps and any metal fittings such as metal sinks. The few electrical fittings which are allowed in bathrooms, such as heated towel rails, will already have their own earth connections. Connections must be made with a proper earthing clamp with a label: 'SAFETY ELECTRICAL CONNECTION – DO NOT REMOVE'.

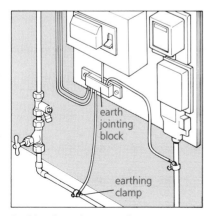

earth jointing block

earthing clamp

Earthing the mains water pipe

Draining the cold water system

There will be occasions when you want to drain down the cold water system. (For details on how to drain down the central heating system, see Chapter 13.)

There may be two drain taps in an indirect system: one positioned just above the mains stopcock and one on the feed to the hot water cylinder. The first is not a lot of use in an indirect system, since once the mains stopcock has been turned off, most of the water left in the rising main can be drained out from the kitchen tap. There will be a little left in the last bit of pipe, but you only want to remove it if you are working on this length of pipe. The other is more useful and is, in fact, the only way of draining the hot water cylinder.

The cold water cistern is drained by shutting off the mains stopcock (or, if fitted, the servicing valve) and then opening the bath and basin cold taps. If you just want to work on the length of pipe between the cistern and the taps, shut the gatevalve first (if there is one) – opening the taps will drain the pipe.

If the mains stopcock won't shut (and there is no servicing valve), the water supply to the cistern can be stopped by tying up the ballvalve. The simplest method of doing this is to remove the lid of the cistern (if it has one), laying a length of wood across the cistern and tying a piece of string to the ballvalve arm so that it is held closed. (If you ever have to do this, leave the piece of wood with its string attached next to the cistern so that it can be used again.) This is usually a better solution for draining the cistern as it leaves a cold supply at the kitchen sink.

Corrosion and dezincification

If two different metals, such as copper and zinc (galvanised cisterns are zinc coated), are in the same system corrosion will occur. With copper pipework, a small amount of the

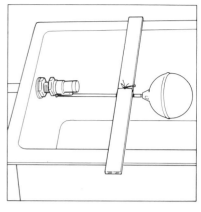

Tying up the ballvalve to drain the cistern

copper will be dissolved in the water (especially if copper filings are left in the system) and this copper will be deposited on the galvanised surface. Each little bit of copper and the zinc with which it is in contact act like a small battery: minute electrical currents are produced, with the water acting as the connection between the copper and zinc to complete the 'circuit'. This current results in small amounts of the zinc being dissolved – the zinc (galvanised) coating becomes porous and the water comes into contact with the steel underneath, which will start to corrode.

The traditional way to prevent corrosion is to suspend a 'sacrificial' magnesium anode in the cistern; the best solution with a corroded cold water cistern is to replace it with a new plastic one.

In some areas, there will be a reaction between chemicals that occur naturally in the water and the metal used in some brass fittings. This reaction – known as **dezincification** – dissolves the zinc from the brass (brass is an alloy of zinc and copper), leaving them as porous copper, which can then disintegrate.

The effect of dezincification can be avoided by using gunmetal or special inhibited brass fittings. Your local water company will be able to tell you whether your area requires any of these measures.

Water softening

Most people are aware of the difference between 'hard' water, which contains dissolved minerals, and 'soft' water which doesn't. Not only do they taste different, but they are very different in their performance when it comes to washing. The minerals in hard water combine with soap to form a 'scum' which can leave a ring around baths and basins. It's only when all the minerals have been 'mopped up' by the soap that it can form a lather.

Some of the dissolved minerals are 'temporary' hardeners and will be removed if the water is heated above 70°C, but this will give rise to fur or scale, not just in kettles but in other hidden parts of the hot water system. More permanent hardness can be removed in one of two ways. The first is simply to add a softener, such as 'Calgon', to the washing machine at the same time as adding detergent, although many detergents contain softeners these days. The second is to install a water softener into the cold water supply. A water softener must be fitted *after* the supply to the drinking water so that this is left providing hard water for drinking.

A water softener works on the principle of 'ion exchange'. It will need *regenerating* every so often. This simply involves the addition of some household salt either manually or automatically.

Lifting floorboards

Lifting floorboards isn't always easy. If they are tongued-and-grooved, the tongue will have to be cut off at least one floorboard: a floorboard saw or circular saw can be used for this. Just to be safe, turn off the electricity and remove the fuses for the downstairs lighting circuit and the upstairs ring main while you do this on the first floor – these cables are under the floorboards.

Floorboards tend to be fitted in long lengths and taking up a whole length is an unnecessary bore. To cut through a floorboard to give a short piece to raise, it must first be lifted up from its neighbours. If you don't do this, the adjacent floorboards will be damaged even if you use a floorboard saw. A good tool for lifting floorboards is a wide bolster chisel. Make the cut through the floorboard as near as possible to the middle of a joist. The nails indicate the joist positions and you should aim to cut as close to the nails as possible without damaging the saw teeth, leaving the nails securing the piece you are not removing.

Once the floorboard piece has been cut through and removed, you will be able to see whether any electric cables and/or central heating pipes are likely to cause trouble when you come to put the floorboards back down again. If there are no electric cables, you can replace the fuses without worry. It is a good idea to paint floorboards to mark the position of pipes and cables.

Fitting the water softener is a matter of breaking into the rising main and 'teeing' one pipe off to the softener and another from it; a third pipe connecting the two tees contains a by-pass valve which is shut when the water softener is in use and open when it is not. Servicing valves should be fitted in both the inlet and outlet pipes of the softener so that it can be removed or isolated for maintenance without the need to turn off the mains water supply.

The softener will need to be connected to a waste and overflow pipe leading to the drains and will require its own electrical connection from a fused connection unit for the clock.

Many water companies require a non-return valve to be fitted in the inlet pipe to the softener and, where the pressure is very high, a pressure reducer in the rising main.

Other water treatment

There are three other devices you might want to fit: a water conditioner, a scale reducer or a water filter.

A **water conditioner** is fitted to the rising main just above the mains stopcock. It does not affect the chemical composition of the water, but rather affects water condition (typically by a magnetic field) so that the magnesium and calcium ions are no longer precipitated.

A **scale reducer** contains polyphosphate crystals which modify the hardness salts so that they do not form scale. It is usually fitted on branch pipes leading to electric showers or gas multi-point heaters (or combination boilers) and has two isolating valves.

A **water filter** is fitted to the branch pipe supplying the sink and connected to a separate tap. 'Kits' are available with self-cutting valves, which can be fitted without turning off the water supply.

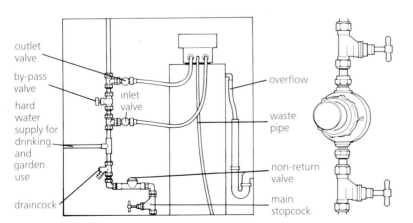

Left: fitting a water softener
Right: a water conditioner is fitted just above the stopcock in the rising main

outlet valve
by-pass valve
hard water supply for drinking and garden use
inlet valve
draincock
overflow
waste pipe
non-return valve
main stopcock

Replacing the cold water system

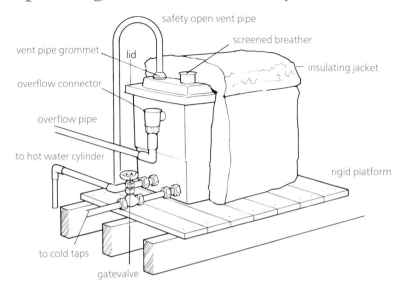

safety open vent pipe

screened breather

vent pipe grommet

lid

insulating jacket

overflow connector

overflow pipe

to hot water cylinder

rigid platform

to cold taps

gatevalve

You may want to replace the cold water system either to convert a direct system into an indirect one or to replace old lead piping (or a rusting galvanised cistern) with more modern copper or plastic.

Installing a new water system is not a difficult job, but it takes time. You can also be without a water supply while you do it. The trick is to get as much of the system installed as you can *alongside* the existing system so that you can change over all in one go.

The two main jobs are replacing the rising main and installing a new cold water cistern together with new pipes leading to the various fittings around the house.

Replacing the rising main

If you have a lead rising main, it may well be that you can see it only at the mains stopcock, after which it disappears into the wall to re-emerge in the loft covered with sacking on its way to the cold water cistern.

The first decision is where to site the new rising main. There are two considerations: first, it must be positioned so that it can be connected easily to the existing stopcock and to the branch which feeds the drinking water tap in the kitchen; second, it must be positioned so that it will not be subject to frost damage. You will also want to place it so that it is not unsightly. Generally, it will be best to run it up the corner of the kitchen which coincides with the corner of a room upstairs. Which corner you choose may also depend on where you want to position the cold water cistern. Although it is no problem running pipes within the loft, it is a good idea to keep all pipe runs as short as possible.

Since you are not connecting up at this stage, the rising main can be installed at a leisurely pace. Use 15mm copper (or plastic) pipe in as long lengths as you can carry home. All connections in the pipe must be accessible so the fewer there are, the easier it is to cover up the pipework. Wherever copper pipework is to show, try to use capillary fittings as they are much neater.

Downstairs pipework The first piece of pipework will need a tee for the branch pipe to the kitchen tap which may itself need tees for pipes to a garden tap and washing machine. Fit these pipes as you go, including the necessary servicing valves.

Passing through the ceiling You will need to make a hole in the kitchen ceiling and the floorboards above for the rising main to pass through. Take up the carpet upstairs and lift the relevant floorboard to see where and how the hole should be made.

You will now be able to see the type of ceiling which is fitted below the joists – either plasterboard or lath-and-plaster. Make a small hole through the ceiling next to a joist with a bradawl, leaving it in place while you go downstairs and transfer the position (and width) of the joist on to the ceiling downstairs with pencil marks. This will now tell you where to make the hole without attempting to drill through a joist.

Treat the upstairs ceiling (loft floor) in a similar way to make the hole to pass up the pipe. Once you are in the loft, it is better to use compression or push-fit fittings since they allow for some adjustment later and the fire risk of using a blowlamp is avoided.

Connecting up When the new rising main is in place (together with a new cistern and the pipes leading from it, if you are doing this), you can connect up to the mains stopcock and to the cold water cistern. You may need a special connector to join to an old mains stopcock (see page 29): fit a drainvalve immediately above this. The connection to the cistern ballvalve is made with a ½in tap connector: fit a servicing valve just before this.

Flush through the pipe before making the final connections and fit insulation to all exposed pipes afterwards.

You can leave buried lead pipe in place. Any that is exposed can be taken out and sold for scrap.

Installing a new cistern

The step-by-step guide opposite gives the main stages in installing a cold water cistern.

Water supply If you're simply replacing the old cistern – and not putting in a new system alongside the old – you will be without water for some considerable time while the job is done. Make sure there are sufficient buckets, bowls and jugs for cooking, drinking, washing and WC flushing.

Working in the loft Working in the loft is unpleasant and somewhat hazardous. Wear old clothes – and a simple face mask – if the loft is very dirty. Make sure there is adequate light (take a wander light up if there is no loft light) and use boards to stand on to avoid crashing through the ceiling. Gardener's knee pads are a good idea: kneeling on joists can be very uncomfortable.

Removing the old cistern Getting the old cistern out can sometimes be a problem – particularly if it's been there for some time. The first difficulty (after baling out the bottom 50mm of water) is removing the old pipe connections. Old galvanised iron rising main and cold feed pipes can usually be removed with Stillsons; a metal overflow pipe may be so rusted that the only answer is to cut it off. Lead pipes are easy to cut through with a hacksaw.

Taking out the old cistern itself can be even more difficult since they were usually put in place as the house was built, so will not go through the loft hatch. If you're desperate to get rid of it, use a general-purpose saw to cut it in small sections (very hard and dirty work!) so that it can be passed down through the loft hatch; otherwise leave it in an unused corner of the loft. If you are cutting it up, work on a large plastic sheet to catch all the filth from inside.

Platform A plastic cistern must be well supported: 230 litres of water weighs a lot and the cistern should be placed over an internal loadbearing wall for extra support. A proper timber platform is necessary for a plastic cistern.

Positioning The cistern must be positioned away from draughts and dirty locations and must be accessible for maintenance (at least 350mm clearance above). Pipe runs in the loft should be kept to a minimum.

Making holes Holes need to be made for all the connections: ballvalve, cold water feeds and overflow pipe. These will probably all be slighty different sizes (typically the holes needed for ballvalves are 21mm, tank connectors 30mm and overflow pipes 25mm) and it is important to use the correct size of holesaw or tank cutter to make them. Make sure that the holes themselves are smooth by using a file and/or trimming knife to finish them off. The position of the hole for the ballvalve stem may be dictated by a backplate or specially strengthened section near the top of the cistern. Holes also need to be made in the lid for the vent pipe grommet and screened breather.

If you're putting in new pipework, it might be easier to make all the holes before the cistern is taken up to the loft; with a replacement cistern, it is important to make the holes in the correct position to take the existing pipes. The holes for the cold feeds must be at least 30mm above the base of the cistern to prevent any dirt or debris in the bottom of the cistern being taken down the pipes. Make the hole for the supply to the hot water cylinder higher than the cold supply – so that if the water fails, the hot runs out first.

The overflow pipe (which should be 22mm in diameter) should be positioned such that the top is at least 20mm below the discharge from the ballvalve and the bottom is at least 25mm above the waterline.

Fitting the connectors If you're installing a new cistern, you can fit the connectors before taking the cistern up to the loft. The ballvalve stem may have both metal and plastic washers each side (plastic closest to the cistern): the tank connectors for the cold feeds and overflow pipes have plastic washers only. Do not use jointing compound on plastic cisterns – it can soften and crack them.

To prevent damage, unscrew the ball from the end of the float arm.

Connecting up The connections to the cistern are straightforward: the ballvalve is connected via a tap connector, which has a small fibre or rubber washer to make the seal (a bent tap connector is often best for this job); the cold feeds are connected to the tank connectors with compression joints and must fall gently away from the cistern to avoid airlocks; the overflow pipe can be copper or plastic (with a plastic tank connector) and should be installed with a fall of at least 1 in 10 from the cistern. The overflow connection should have a filter to keep out insects and an internal dip tube to prevent icy draughts. The overflow itself should discharge in an obvious place where it cannot be ignored.

Don't forget to fit a servicing valve just before the ballvalve and gatevalves to all the cold water feeds: these are connected after a short run of pipe except for the feed to the hot water cylinder, where the gatevalve is usually fitted in the airing cupboard near to the cylinder.

Fitting gatevalves means that new pipes to fittings around the house can be installed *after* the cold water cistern is in place and connected up. Once the servicing valve is fitted the water can be turned back on.

It sometimes helps, once the ballvalve has been fitted, to run a little water into the bottom of the cistern (below the outlet points) so as to anchor the cistern in position while you work on it. Clean it out first.

The vent pipe from the hot water cylinder (22mm) should rise at least 400mm above the water level in the cistern before turning down over the cistern. It should pass through a grommet in the cover and terminate about 50mm above the water level.

Filling and testing Before filling the cistern, it should be cleaned thoroughly and sterilised with a dilute bleach solution (or use Milton). Check carefully for leaks as the cistern is filled and adjust the float arm so the water level is 50mm below the overflow level.

Pipe support When installed, the pipes must not put any strain on the tank (particularly a flexible plastic one). Clip them securely to the joists and to the cistern platform; this will also reduce noise from the system.

Label all gatevalves so you know which is which in an emergency.

Insulation The cistern will need to be fitted with a rigid, close-fitting and securely fixed cover which excludes light and insects but which is not airtight. In practice, this can be achieved with a screened breather fitted in the cover. The cistern must also be well insulated (as will all pipes, including the open vent).

Bye-law 30 kits

These have all the 'extras' necessary for a cistern to meet the requirements of the water bye-laws. A typical kit includes:
- a cover
- an insulating jacket
- a screened breather
- a vent pipe grommet
- an overflow pipe connector with insect filter and internal dip tube
- a metal ballvalve support plate.

Installing a new cold water cistern

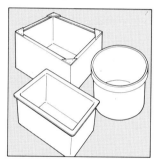

The three main types of cistern: galvanised iron *top* rigid plastic *bottom* flexible plastic *right*

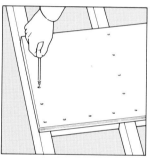

Make a secure platform to support the new cistern (full weight 237kg). Use 22mm chipboard or plywood

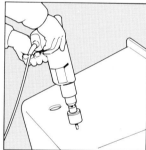

Make correctly sized and positioned holes in the cistern for the ballvalve, the overflow pipe and the cold water outlets

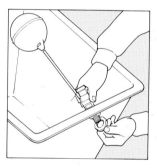

Connect the ballvalve to the cistern with its two nuts and washers and fit a tap connector for the rising main connection

Fit tank connectors for the cold feeds (fit plastic washers both sides) and connect a short length of pipe leading to a gatevalve

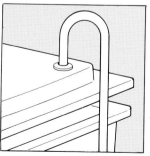

The vent pipe from the hot water cylinder is fitted over the top of the cistern and passes through the grommet in the lid

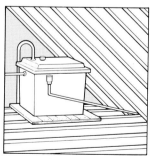

Fit the overflow pipe using a suitable connector and take the sloping pipe out through the eaves at a suitable point.

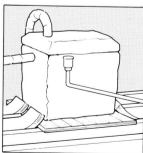

After cleaning and testing, fit the insulating jacket around the sides and over the top. Don't insulate the loft floor underneath

Rewashering a ballvalve

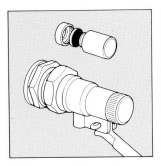

A Portsmouth ballvalve has retaining caps to hold in both the piston and, usually, the washer inside the piston

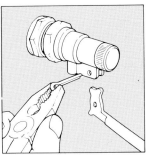

After turning off the water, use pliers to extract the split pin. Flatten the open ends, pull the pin out and remove the float arm

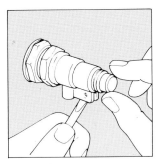

Unscrew the end cap and use a screwdriver to lever out the piston assembly. Catch it before it falls in the water

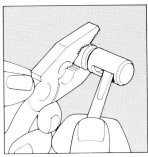

Unscrew the end cap using pliers and remove the washer. Clean up the valve parts and reassemble with a new washer

Repairing a burst pipe

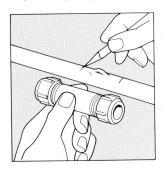

Turn off the water and mark the section of pipe to be cut out, taking the length from the connector

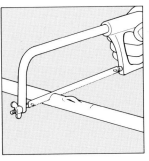

Cut the pipe and clean up the ends of pipe inside and out. Be careful that no copper filings are left inside the pipe

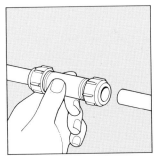

Slide the slip connector over the ends of the pipe and centre it. Don't use jointing compound unless recommended

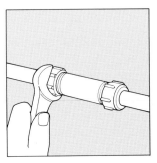

Tighten the compression joints using a spanner at either end. Turn the water on again and check for leaks

Running new pipes

All connections to the cold water cistern should start with a gatevalve so that the new section can be isolated if necessary

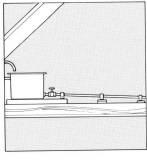

Always run pipes with a fall away from the cistern and make sure that they are well supported with pipe clips

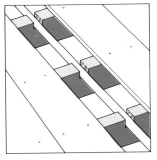

If you have to cut a notch in a joist to run a pipe, cut it in the middle of the floorboard and leave a little room for expansion

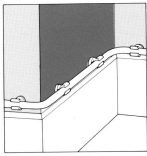

Where pipes turn corners with elbows or bends, don't put them hard up against a wall. They need room to expand

CHAPTER FOUR

THE HOT WATER SYSTEM

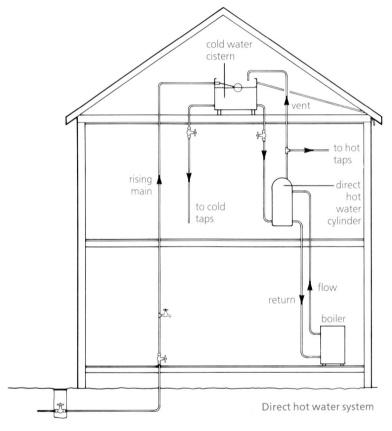

Direct hot water system

Storage systems

There are three main kinds of hot water storage heating system: in the first two, the water is heated directly or indirectly by some kind of boiler; in the third, it is heated by an electric immersion heater.

Direct boiler systems

In a **direct** boiler system, the copper hot water cylinder (or, in older installations, galvanised hot water tank) is supplied with water from the cold water storage cistern; the water then passes down to a heating boiler, back boiler (behind a fire) or gas circulator to be heated before returning to the hot water cylinder. A pipe leading from the top of the cylinder or tank then takes the hot water to the rest of the house. The pipe bringing the water from the boiler (the *flow*) is connected near the top of the cylinder; the pipe taking water back to the boiler to be heated (the *return*) is connected near the bottom. (In a galvanised tank, both pipes are connected to the bottom, the flow continuing up inside the tank to discharge near the top.)

The whole system usually depends on gravity – which means the cold water storage cistern must be above the highest point of the hot water cylinder (usually the hot water draw-off), and the hot water cylinder must

In the majority of houses there are two different types of hot water: one which is in continuous circulation around the radiators of the central heating system, and one which comes out of the hot water taps in the bathroom and kitchen. The second type is the subject of this chapter – it is often known as the *domestic hot water* system.

There are many different ways of heating the domestic hot water. Broadly, they break down into two main types: one where the water is heated and *stored* hot in a hot water cylinder or tank; the other where cold water is heated *instantaneously* as it is drawn off.

be above the boiler so that the heavier (cooler) water falls and the lighter (heated) water rises through the various connecting pipes. If boiler and hot water cylinder are on the same level, a **pump** has to be installed into the system. Sometimes, a **safety valve** may be fitted into the boiler flow pipe; a **draincock** is always fitted in the return. A vent pipe must be fitted to the top of the cylinder to terminate over the cold

water cistern and the hot water draw-off is taken from this *above* the level of the top of the cylinder.

Pipe sizing is important, too. The cold feed to the cylinder and the vent pipe must be at least 22mm in diameter; it is best if the hot water distribution pipes are 22mm as well – at least as far as the connection to the branch pipe for the bath. The pipes leading to and from the boiler are usually 28mm (or even 35mm).

Direct hot water systems all suffer from one major drawback, which is the formation of **scale** in the boiler pipes and in the cylinder, particularly in hard water areas – see below. They are no longer installed.

Indirect boiler systems

The difference between a direct hot water cylinder and an indirect one is that the latter has a heat exchanger, often in the form of a coiled pipe, which takes the hot water from the boiler and passes its heat to the domestic hot water in the cylinder, the two never mixing. The water in the coil, now cooler, is passed back to the boiler. The *flow* connection will be towards the top of the cylinder and the *return* connection towards the bottom. With older, annular-type heaters, the tappings can be on opposite sides of the cylinder. **Draincocks** must be provided for both circuits.

This type of system requires *two* cold water cisterns: the normal one which supplies the cold feed to the hot water cylinder (the 'secondary' circuit); the other a feed-and-expansion cistern for supplying the 'primary' boiler circuit. Each has a vent pipe from its respective circuit. Often, of course, the primary circuit will be part of a central heating system. In older systems, this will usually be a gravity-fed circuit separate from the pumped central heating circuit, but in modern systems it will be a pumped circuit which operates together with the central

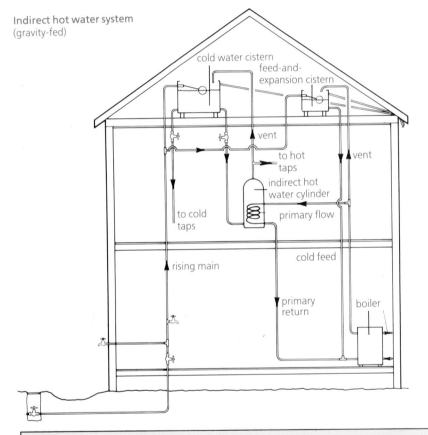

Indirect hot water system (gravity-fed)

cold water cistern
feed-and-expansion cistern
vent
to hot taps
vent
indirect hot water cylinder
to cold taps
primary flow
rising main
cold feed
primary return
boiler

Scale

Water naturally contains dissolved chemicals: the more it has, the *harder* it is. Apart from the problems of washing in this hard water (soap will not lather) when the water is heated, the dissolved chemicals are 'precipitated' out as insoluble carbonates; the higher the temperature, the more the salts that are precipitated. This effect is familiar as the 'furring' inside kettles.

Fur or scale is bad enough in kettles,

but can have a disastrous effect on the insides of pipes and boilers in direct systems. Not only is it a bad conductor of heat – so putting up running costs – but the fur can reduce the effective internal diameter of a pipe and can eventually lead to boilers leaking, immersion heaters burning out, and noises in the water heating system.

Scale formation can be prevented in direct boiler systems by fitting a water

softener, water conditioner or scale reducer (see page 42) or by keeping the water temperature below 60°C.

The best way of preventing scale, however, is to replace the cylinder with an **indirect** one where the same water is circulated round and round the 'primary' circuit (the one which passes through the boiler) which, once it has given up its chemical salts, is in effect softened so that no more will be deposited.

heating. See also Chapter 11.

The level of water in a feed-and-expansion cistern should be low enough to allow the water to expand as it gets hot without overflowing. The vent pipe is often wrongly called an expansion pipe: its purpose is to allow any air and steam to escape safely (the expansion actually takes place up the cold feed pipe).

In addition to the traditional vented system, it is now permissible to have an *unvented* hot water storage system, which does away with the cold water cistern altogether and has other advantages (see page 58 for details); sealed central heating systems (which do away with the feed-and-expansion cistern) have been legal for some time.

Immersion heaters

Another way to heat water is to have an electric immersion heater which fits into a 'boss' on the hot water cylinder. The immersion heater can be mounted vertically or horizontally (depending on the design of the cylinder); it may be possible to fit two heaters or a dual-element heater so that you can choose whether to heat the whole cylinder (for a bath) or just the top part (for washing up and hand washing). A **timer** can be fitted to an immersion heater so that it comes on at specified times of the day. If the cylinder is well insulated, advantage can be taken of the cheaper Economy 7 tariff. The water is heated at night and, if not used, stays hot during the day. You need a special electricity meter for this.

Immersion heaters can be fitted instead of or as well as a boiler to heat the water. A common arrangement is to have both so that the immersion can be used as a back up and in summer when the boiler is off.

Where electric heating is the only source of hot water, an alternative to a hot water cylinder is an electric water heater, wall-mounted or fitted in a cupboard.

Instantaneous systems

Most people are familiar with the 'Ascot' type of **gas** instantaneous single-point water heater fitted in the kitchen for supplying hot water to the sink. This method of heating has the advantage of being more economical as no heat is wasted keeping water hot and none is lost in the pipes between the hot water cylinder and the hot tap. Is is not unusual to have a sink water heater even where the rest of the house is supplied from a hot water storage cylinder.

In addition to this type of gas water heater, there are **electric** instantaneous water heaters for providing hot water for showers.

Both gas and electric instantaneous heaters are used for supplying water at a single point. For supplying hot water for a whole house, a gas **multi-point** heater can be used. This operates on the same principle as a single-point instantaneous heater, but is more powerful and can be linked to several outlets. It can be connected to the rising main and is often fitted in houses with direct cold water systems to provide the hot water. It can be used to provide the hot water supply for a washing machine, which is not the case for a single point heater. A **combination boiler** combines the functions of a central heating boiler

and a multi-point water heater.

The main disadvantages of an instantaneous water heater compared with storage systems is that the flow of water is often slower and, in the winter when the water in the mains is colder, the hot water will be less hot. They also suffer from scale formation in hard water areas. On the other hand, hot water is always available with this type of heater.

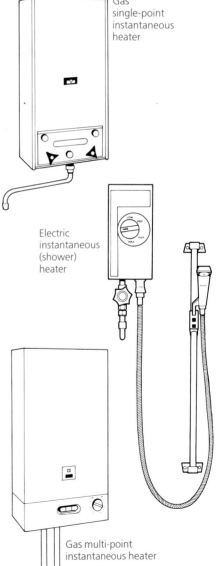

Gas single-point instantaneous heater

Electric instantaneous (shower) heater

Gas multi-point instantaneous heater

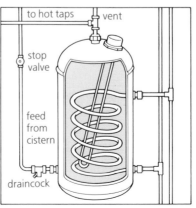

Immersion heater (with indirect cylinder)

Designing hot water systems

A system which uses only an immersion heater to provide the hot water is the easiest to design and install. The important points to remember are that pipe runs from the hot water cylinder should be kept as short and as direct as possible to avoid heat loss – and a long wait at the tap before hot water arrives. The maximum recommended length for these uninsulated 'dead legs' is 12m (39 ft): it is better to insulate all hot water pipes wherever possible.

Pipes must be installed with a fall away from the take-off point (in the vent pipe above the cylinder) to prevent *airlocks* forming.

Instantaneous water heaters do not pose many design problems either. The important things here are that a gas water heater may require its own flue (usually a 'balanced flue' – see Chapter 11) and that an electric water heater will require its own electric circuit (see Chapter 8). Single-point instantaneous heaters will be fitted as near as possible to the sink or basin that they serve: multi-point heaters can usually be positioned so that pipe runs can be kept to a minimum. This assumes that the bathroom is next to or above the kitchen (as is usually the case); if running hot water to another room (a bedroom, say) will create a very long pipe run, it might be simpler to install a separate instantaneous or small storage heater.

All instantaneous heaters are run from the rising main.

It's when you come to systems which run from some kind of boiler that the design needs more consideration. The main points to think about are the positioning of the boiler and hot water cylinder, the sizes and running of pipes, the types of pipe fittings to be used and the avoidance of airlocks. You will probably also want to fit some kind of thermostatic control.

Positioning

The positioning of the boiler may well be the first consideration. If it is a boiler that is also to be used for central heating, the best position will probably be in the kitchen on an outside wall. Some existing 'boilers' used for hot water heating will, however, be *back boilers* which are situated behind gas or solid fuel fires in the main living room, which may mean significantly long – and inefficient – pipe runs. The hot water cylinder should certainly be as near as possible to the boiler – preferably vertically above it – and to the taps it serves. With a back boiler, this could mean putting the hot water cylinder in a bedroom airing cupboard.

Pipes and fittings

The correct sizing of pipes is important for hot water systems – particularly for the pipes between the boiler and hot water cylinder in a gravity system. To reduce pressure lost in going around corners, pipes should be bent or fitted with slow bends rather than elbow fittings.

No valves should be fitted to obstruct the cold feed pipe to or the open vent pipe from the boiler – apart from a draincock at the lowest point of the system. No valves should be fitted in the vent pipe leading to the cold water storage cistern, but it helps for maintenance if servicing valves are fitted in the hot water pipes.

Airlocks

An 'airlock' occurs when air gets into a pipe and prevents the flow of water. The flow from the tap stops – and there may be hissing or bubbling noises as well.

Airlocks happen in hot water systems for a number of reasons. It may be that the pipework does not slope gently away from the vent pipe which can cause air to collect at the highest point of the pipe. It may be that the cold pipe to the hot water cylinder is too small and cannot fill the cylinder fast enough when the bath tap is being run, with the result that the water level in the vent pipe falls and allows air into the system. Another possible cause is that the cold water cistern is too small or that the ballvalve supplying the cistern is blocked or slow in operation. Partly-closed valves (or the use of stop-valves) can also cause airlocks.

Whatever the cause of an airlock, a temporary cure can usually be effected by connecting a hose pipe from the kitchen *cold* tap (under mains pressure) to the tap which is not running and turning both on. The water being driven along the hot pipes should get rid of the bubble of air. In practice, running a hose pipe all the way from the kitchen sink to (say) the bath hot tap (to say nothing of the problem of connecting it at both ends) may not be practicable and you should first try using the nearest cold tap in the same way. An alternative method is to attach an electric drill water pump attachment to the tap.

If a design problem is the cause of the airlock, this procedure will have to be repeated every time, so it will be better to find the cause and remedy it by fitting bigger pipes or a bigger cistern (or a second cistern alongside the first – see Chapter 3), or by repositioning the pipes so that they have the correct fall.

Curing an airlock

Thermostatic control

An immersion heater has the simplest type of thermostatic control – fitted in the centre of the heater and adjustable by the user.

A solid-fuel boiler, whether a separate boiler, an Aga-type cooker/hot water boiler or a back boiler, is the most difficult to control and will heat the water to some extent all the time that the fire is lit. The hot water cylinder is, in fact, used as the 'heat sink' for many solid fuel boiler systems allowing somewhere for excess heat to be dissipated (sometimes a bathroom towel rail is used for the same purpose). Great care should be taken before attempting to drain the water out of this type of system – the boiler must be let out first.

Systems which operate on the type of boiler which can be turned on or off (gas-fired or oil-fired) can have one of a number of controls fitted. A boiler has its own thermostat to control the water temperature. In addition, a **cylinder thermostat** can be strapped to the cylinder either to turn the boiler off or to operate a valve which shuts off the flow from the boiler to the cylinder when the stored water is hot enough. In a pumped system, the thermostat may also switch off the pump. Where the water is heated by pumped circulation operating in conjunction with a central heating system, the valve may be a *three-way* one which will divert the water either to the hot water cylinder or to the central heating radiators or, sometimes, to both – see Chapter 11.

Another type of hot water control is a **non-electric thermostatic** valve fitted into the return pipe from the cylinder to the boiler. This can be adjusted to close off the flow when the temperature is sufficient.

All these measures will save fuel and many thermostats can also be used in conjunction with a **programmer** to bring the hot water heating on only at certain times of the day.

Hot water cylinders

A hot water cylinder is traditionally put in a cupboard which also serves for airing clothes. In many modern central heating systems, the airing cupboard may also be the place where much of the control gear is situated since this is convenient for access and for arranging the pipework.

The standard size for a hot water cylinder is 900mm (36in) high and 450mm (18in) diameter. This gives a capacity of around 120 litres (26 gallons) which should be sufficient for one full bath. The next most common size is the same diameter but 1050mm (42in) high, with a capacity of 140 litres (31 gallons). If you go to a specialist shop, such as a plumbers' merchant or central heating specialist, you will be able to find other sizes. A tall thin cylinder (300mm x 1600mm, for example) is particularly useful when replacing an old galvanised hot water tank which has a cupboard built to fit it. Look for cylinders made to BS1566.

Installing a hot water cylinder

The most likely reason for fitting a new copper hot water cylinder is that you want to replace an existing galvanised hot water tank. If you simply want to change a direct cylinder to an indirect one or to fit a cylinder with an immersion heater boss, there are ways of adapting an existing direct cylinder – see *Upgrading cylinders* (page 52). If you are replacing one copper cylinder with another, you may find one with the connections in the same place. (Step-by-step drawings are on page 54.)

First buy all the fittings you need. The job will take some hours and you will be without hot water while you are doing it, so you need to have everything to hand before starting. The old tank can usually be drained from the draincock at the boiler or gas circulator (after turning this off):

make sure that the gatevalve on the cold water supply to the hot water tank is closed before you do this. If there is no gatevalve in this supply, the cold water cistern will have to be drained: turn off the supply or tie up the ballvalve (so that there is still a supply to the kitchen tap) and open the upstairs cold water taps. It would be a good opportunity to fit a gatevalve to this supply and, if the supply pipe is made from lead, to replace it with copper or plastic.

The most difficult part of getting out a galvanised hot water tank is likely to be disconnecting the pipes leading to and from the boiler – particularly if they have been there for some time. The largest pair of Stillsons may shift the nuts, but you may have to resort to cutting through the pipes.

There will be some water and debris in the bottom of the hot water tank – so make sure that buckets and cloths are to hand to catch the mess.

Once the old tank has been removed, you can install the pipework for the new cylinder. The pipes you are likely to need are:

☐ 22mm cold feed from the cold water cistern
☐ 28mm pipes leading to and from the boiler for gravity circulation, or 22mm if connected to the pumped circuit for pumped circulation
☐ 22mm vent pipe leading up to and over the cold water storage cistern
☐ 22mm hot water supply pipe leading off the vent pipe and going first to the bath hot tap

Although the positioning of the pipes and the fittings needed will depend on the exact position of the cylinder and its connections, it will usually be easier to get the pipes in before the cylinder is installed. If space is very tight, use flexible copper pipe or plastic pipe for some if not all of the

Types of cylinder

When looking in a shop or through a manufacturer's or mail-order catalogue for hot water cylinders, you will find several different types. Special cylinders are needed for *unvented* systems. See page 58.

Direct This type of cylinder is designed for direct boiler systems, so has two tappings to take fittings for the pipes to and from the boiler but no internal heating coil. Existing direct cylinders can be *upgraded* to indirect – see below.

Indirect This also has two tappings, but these will be connected internally to a heating coil – see drawing on page 49.

Both indirect and direct cylinders are usually fitted with a 2¼in BSP boss for a vertical or horizontal immersion heater. *High-recovery* cylinders heat up in half the time so can be smaller in size.

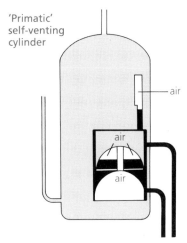

'Primatic' self-venting cylinder

air

air

air

Combination cylinder (direct)

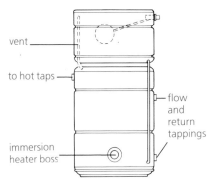

vent

to hot taps

flow and return tappings

immersion heater boss

Pre-insulated These cylinders are already covered with a layer of insulating foam which does away with the need for adding a jacket. Generally the extra cost for this type is around the same as for adding your own jacket – but the insulation won't fall off.

Combination Sometimes referred to as a 'packaged plumbing' unit, a combination cylinder has its own cold water storage cistern fitted on top of the cylinder. This arrangement can be used where there is no room for a separate cold water storage cistern in a flat or a house with a flat roof. Combination cylinders come in different sizes and are available for both direct and indirect boiler heating systems – the indirect version may have *two* cold water cisterns, one to feed the cylinder and the other as a feed-and-expansion cistern for the central heating system.

Self-venting This type of cylinder (also called 'Primatic') does not require a feed-and-expansion cistern for the primary circuit to the boiler. Instead, the mechanism inside

the cylinder allows the 'primary' circuit to be filled up from the cold water supply to the hot water cylinder. When it is full, a large air bubble prevents the two mixing – provided the water in the boiler circuit is never allowed to boil. Self-venting cylinders cannot be used with corrosion inhibitors and are seldom installed these days.

Packaged If you are installing central heating, you can buy pre-insulated cylinders with various central heating components (such as the pump, motorised valve, thermostat and programmer) already pre-plumbed and pre-wired in place.

Economy 7 An Economy 7 cylinder is a large capacity (typically 210 litres) pre-insulated cylinder fitted with two horizontal immersion heaters. The heater at the bottom is wired so that it will operate on the cheaper tariff at night to heat up the whole cylinder; the heater at the top can be used during the day (on full-rate tariff) for 'topping up' when necessary.

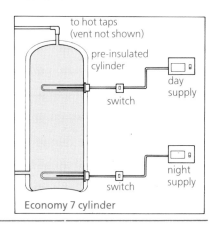

to hot taps (vent not shown)

pre-insulated cylinder

day supply

switch

night supply

switch

Economy 7 cylinder

final fittings. See also Chapter 13.

The fittings provided on most cylinders are 1in male BSP for the boiler connections and 1in (or ¾in) female BSP for the cold inlet and hot outlet. These should all be fitted with parallel-threaded (*not* tapered) iron-to-copper fittings so that the copper pipes can be connected directly. The threads of the male iron parts of these fittings should all be wrapped with PTFE tape to ensure a watertight seal. The cold inlet at the bottom of the cylinder can be fitted with a male iron to copper elbow containing a draincock.

The cylinder itself should be positioned on a flat rigid surface: if the floorboards in the airing cupboard are at all uneven or if you need space for access to pipes, it is a good idea to mount it on a small platform.

Compression or union capillary fittings should be used for the final connections to the cylinder so that it can be easily removed for maintenance. Any tappings that aren't being used on the cylinder can be fitted with screwed blanks. Lastly, fit an immersion heater and an insulating jacket (if required).

Before filling the cylinder, open

the hot water taps to avoid airlocks forming and then fill slowly, checking for leaks. If the primary circuit is also part of a central heating circuit, this will also need filling and some of the radiators may need bleeding afterwards to let out air.

Upgrading cylinders

Direct cylinders can be converted simply to indirect ones using a cylinder **conversion kit**. There are two types: one uses the immersion heater boss for both the primary flow and return; the other requires the making of two holes in the side of the cylinder into which a coil can be positioned. The second involves more work but it means that an immersion heater can also be fitted, provided the coil is carefully positioned – see drawings on page 54.

The kit for doing the job comes with the necessary fittings so that the job of 'winding' the coil into one hole and out of the other after removing the pipes from the boiler should take only 15 minutes.

Fitting a new **immersion heater boss** to a cylinder is a longer job. There are different designs of flange for fitting to cylinders, depending on where the flange is to be fitted. The 'Essex' flanges are specially constructed in sections, plus two rubber jointing washers, enabling them to be inserted and fixed from outside the cylinder.

If fitting a horizontal heater, remember that it should neither be positioned below the level of the cold water inlet nor interfere with any indirect coil inside the cylinder. Don't fit an immersion heater boss to an indirect self-priming cylinder.

The hole for the flange can be made with a large tank cutter or by drilling a circle of small holes and using a padsaw to connect them up, preventing the disc falling into the cylinder. The hole will need to be filed smooth and all copper filings washed out of the cylinder before the flange is fitted.

All the jobs above require the cylinder to be at least partly drained. It might be easier to remove it altogether.

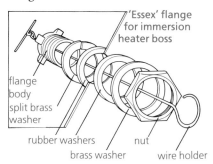

'Essex' flange
for immersion
heater boss

flange
body
split brass
washer

rubber washers   nut
brass washer   wire holder

Immersion heaters

An immersion heater works like the element in an electric kettle. There are three different arrangements for fitting immersion heaters:

☐ a single heater, fitted vertically or horizontally
☐ two heaters fitted horizontally
☐ a dual element heater fitted vertically.

With two heaters or a dual element heater, it is possible to keep a small amount of water hot – enough for hand washing and washing up – whilst a single heater will heat the whole cylinder.

Immersion heaters are generally rated at 3kW and come in different lengths to suit different sizes of cylinder: don't attempt to economise by fitting a shorter vertically mounted one than you need – it will heat only the water above it.

The thermostat, which fits between the element loops, is usually sold separately – so make sure you have got one.

Fitting an immersion heater

The plumbing involved in fitting an immersion heater is relatively minor, provided the cylinder already has an appropriate boss – see left. If the cylinder is already installed, you need only to drain it, unscrew any blanking cap (or old immersion if you're replacing) and screw the new one in place using the washer supplied. You don't need to drain the whole system – just enough so that water doesn't come out of the immersion heater boss. You will, however, need an *immersion heater spanner* for this job (see page 17).

The major part of the job is electrical and if you're not confident about tackling this, get a qualified electrician to do it for you.

To comply with the IEE Wiring Regulations, an immersion heater must have its own electric circuit run from a spare fuseway at the consumer unit, though there are many houses where the supply for the immersion heater has been taken off the upstairs ring circuit.

The new circuit needs to be run in 2.5mm^2 twin-and-earth cable with a 15A (or 20A) fuse or circuit breaker at the consumer unit. The cable is run to a *double-pole* switch (or fused connection unit) from which a length of 15A (1.5mm^2) heat-resisting three-core flex is run to the connections on the immersion heater. Special double-pole switches are made; if the switch can be reached from the bath or the shower, it must be the pull-cord kind.

Two immersion heaters (or a dual-element heater) generally need two flexes. Ideally, these are taken from an *Economy 7 controller*, which automatically brings on the lower (or longer) element at night and provides for a short 'top-up' during the day with the upper (or shorter) element.

There are also special timers for single-element immersion heaters. Some incorporate the necessary double-pole switch; others are simple timers and are fitted between the double-pole switch and the flex to the immersion heater.

Adjusting the heater thermostat

The thermostat of an immersion heater can usually be adjusted by turning off the double-pole switch and removing the heater cover. There is a small slot in the centre of the thermostat which is marked in °C and which can be adjusted with a screwdriver. A temperature of 60°C (140°F) is usually recommended for domestic hot water, and will prevent scale formation in hard water areas. The setting is more precise with horizontal heaters; you may need to experiment with vertical heaters.

Replacing a hot water cylinder

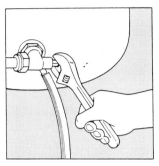

Whichever type of cylinder is being replaced, the first step is to drain down – either at the cylinder or at the boiler (or both)

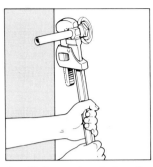

The old pipes to the cylinder will have to be disconnected. This may require force if galvanised iron pipe has been used

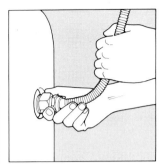

The pipes are fitted to the screwed tappings or inlets on the new cylinder. Use flexible copper or plastic pipe if space is limited

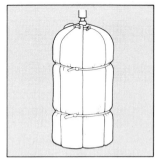

If necessary, insulate the cylinder by fitting a separate jacket, but it is much better to use a pre-insulated cylinder

Fitting an immersion heater

Choose the type of heater which is best for the type of cylinder (or buy heater and cylinder at the same time)

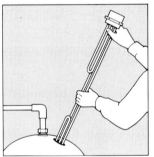

Fit the immersion heater with its washer and insert it into the screwed boss. PTFE tape should not be needed

In order to tighten the immersion heater a special spanner is required. These can be hired but are inexpensive to buy

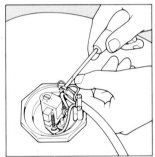

After installing the necessary electrical circuit, connect up the heat-resistant flex to the terminals on the heater

Upgrading a cylinder

Converting direct to indirect

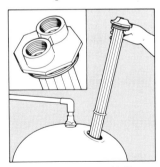

One type of conversion kit fits into the immersion heater boss. It is tightened with an immersion heater spanner (Microversion)

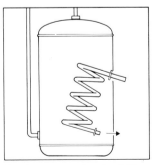

Another type requires the drilling of two holes (using the template supplied) and the 'winding in' of a copper coil (Sidewinder)

Cylinder thermostat

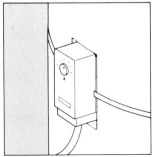

To fit a cylinder thermostat to a pre-insulated cylinder, cut away the insulation to get metal-to-metal contact

Immersion heater timer

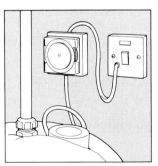

An immersion heater timer is wired in place between the double-pole switch and the immersion heater

A new hot water system

If you are thinking about installing a completely new hot water system there are several considerations:

☐ how much hot water do you need?
☐ is the hot water system to be part of a central heating system?
☐ are there existing flues and/or outside walls to take balanced flues?
☐ do you want a conventional 'vented' system or a modern *unvented* system?

How much hot water?

Any hot water system must meet the requirements of the whole house. Primarily, this is to supply the kitchen sink, bathroom basin and bath plus whatever appliances, such as washing machines, are connected to the hot water supply. The major factor is likely to be the number of baths taken which will vary with the size and age of the family. Reckon each bath at around 60 litres to give you an idea of the amount of hot water storage that is needed. It's worth remembering that showers use much less hot water (see Chapter 8). The *size* of the house may also influence the decision – in a very large house, the length of hot water pipe runs could be unacceptably large for a single system and a combination of storage and instantaneous systems may be necessary.

Hot water and central heating

Few people would think seriously about installing a central heating system these days without also using it to provide domestic hot water. If you're thinking of putting in central heating, the type of system you should install will depend on more factors than the hot water requirements. See Chapter 11 for more details.

Many existing systems which don't already heat the hot water can be adapted to do so, but if yours can't (an early warm air system, say) or if you're installing one of the few types of system that doesn't also heat the hot water (such as electric storage radiators), you will have to choose the hot water system.

The best and cheapest solutions to providing domestic hot water not connected to central heating systems are likely to be an Economy 7 hot water cylinder or, if you have gas laid on, a gas circulator used with an indirect cylinder.

Single gas instantaneous water heaters have their place alongside a storage system; the main problem with relying on a gas multi-point heater to provide all the house's hot water is the low flow rate which means baths take a long time to fill.

Flues

If you use a fuel-burning boiler or water heater to provide your water heating, this must be fitted with a proper flue to take away the products of combustion.

For many boilers, this can be a normal chimney, provided that there is a sufficient supply of *fresh* air into the room for the fuel to burn. For gas multi-points and many wall-hung boilers, a *balanced flue* is used: this is a two-part duct which allows fresh air in and exhaust gases to escape. Note that a single-point gas water heater in a bathroom **must** be fitted with a balanced flue: a good idea for others, too.

Many old gas water heaters do not have proper flues and are dangerous – get the Gas Board to check any you are worried about.

Vented or unvented?

An unvented hot water storage system has many advantages over the normal vented system. See page 58 for details.

Heating by gas

Central heating boilers

These are covered in detail in Chapter 11. Suffice it to say here that all modern central heating boilers can provide domestic hot water via an indirect hot water cylinder or, with a *combination* boiler, instantaneously.

Back boilers

It is unlikely these days that you would want a new back boiler to provide all your central heating needs, but you may well want to replace one which provides domestic hot water.

Gas circulators

A gas circulator is the answer where central heating is provided by a gas warm air system and the only other method of heating the domestic hot water is an electric immersion. To save installing a new Economy 7 hot water cylinder, the existing cylinder can be converted to direct or (better still) indirect operation by a wall-mounted gas circulator. In effect, this is a small boiler which will run more cheaply than an electric one and will reduce dependence on a single fuel.

A back-fitting gas circulator would also be a good choice for replacing an elderly solid fuel back boiler behind a new gas fire.

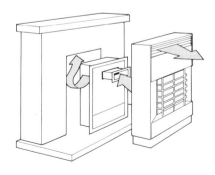

A back-fitting gas circulator

Gas storage water heaters

These are rather like a water heater and storage cylinder in one. The mass of water (typically 90 litres) is heated quickly to a pre-set temperature (60°C) and is maintained at this level. Several sizes are available.

Gas instantaneous heaters

One of the advantages of using an instantaneous gas water heater is that it is much more efficient than using a gas central heating boiler to heat the water in a storage cistern – particularly in the summer. Any gas appliance is more efficient when operating at full load, which is not the case with most central heating boilers when providing only hot water heating.

A typical gas instantaneous water heater will give a temperature rise of 50°C with a flow rate of around 6.5 litres/minute – so if it is taking water from the mains at 10°C, it will raise the temperature to 60°C. The temperature and flow are usually interdependent – the greater the flow rate, the lower the resultant temperature and vice versa – though some models are thermostatically controlled to give a particular temperature whatever the flow rate. This compares with 18 litres a minute or more from a storage system.

The majority of gas heaters will require a conventional flue or a balanced ('room-sealed') flue: in bathrooms, only balanced flues may be used. The pressure 'head' normally needed for operating an instantaneous water heater, typically 10m or 15psi (pounds per square inch), means that they have to be connected to the mains.

Note that it is illegal for persons who are not 'competent' to install gas appliances. Although you can do all the water plumbing side yourself, the gas fitting should be left to a qualified gas fitter who is on the register of the Confederation for the Registration of Gas Installers.

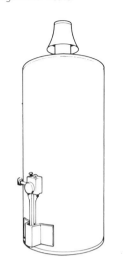

Gas storage water heater

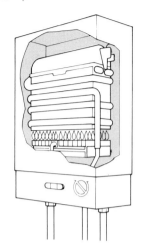

Gas multi-point water heater

Heating by electricity

Electric storage water heaters

A small 'open-outlet' electric storage heater can usefully be positioned over or under the kitchen sink as it will provide sufficient water for washing up. If it is mounted over the sink, a normal swivelling outlet spout is used. For under-sink installation it is essential to use the purpose-designed hot tap supplied by the heater manufacturer. When the tap is turned on, cold water (usually from the mains) is allowed into the bottom of the heater, thus forcing hot water out of the top and

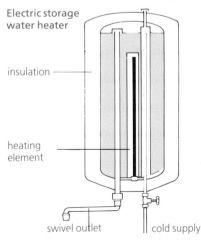

Electric storage water heater

insulation

heating element

swivel outlet   cold supply

out of the nozzle outlet or the hot tap. A anti-drip device inside the heater prevents a nozzle dripping as the water expands in heating. Typical sizes are 7 and 15 litres with 1kW or 3kW loading; some models have adjustable thermostatic controls.

Unvented undersink electric water heaters (typically 10 litres/3kW) use normal taps and can feed more than one outlet.

Larger electric storage heaters will heat 91 or 136 litres (20 or 30 gallons) and replace a conventional hot water storage cylinder. They need to be fed from (and vented to) a cold water storage cistern unless they have their own integral cold water cistern. See also *Economy 7* cylinders on page 52.

Electric instantaneous heaters

Small, 3kW to 6kW, instantaneous electric water heaters, similar to the type used for electric showers, can be used over a sink or basin.

You may also find instantaneous water heaters, rated at 9kW or more, which can feed more than one outlet – ideal for use in a loft conversion, say.

Solar water heating

Harnessing the natural radiant energy of the sun seems an obvious way to cut down on ever-increasing fuel bills. Unfortunately, the majority of energy used in homes is for *space* heating (heating the rooms and the people) and this is mostly needed in the winter when there is not sufficient sunshine in this country for solar energy to be feasible.

Solar heating does come into its own for the fairly specialised use of heating water for swimming pools and can make a contribution to heating the domestic hot water.

So far, the costs of having a solar heating system installed have outweighed the likely savings in fuel bills, but if you can install a system yourself, you should save money.

On average, the amount of solar radiation falling on each square metre of the United Kingdom is from 800kWh a year in Scotland to 1,100 kWh a year in the south west of England. A solar collector area of around 4m² to 5m² is usually considered the optimum.

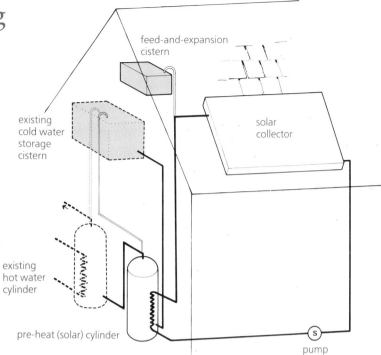

feed-and-expansion cistern

existing cold water storage cistern

solar collector

existing hot water cylinder

pre-heat (solar) cylinder

S

pump

How it works

A solar heating system consists of three main components: the solar panel, a feed-and-expansion cistern (which must be positioned at the highest point of the system) and the solar cylinder. Water from the cylinder passes through the solar panel where it is heated by the sun's rays. This is rather like a radiator working in reverse (in fact, a radiator *could* be used as a solar panel). The water then passes to a coil in the indirect solar cylinder, and is then pumped back to the bottom of the solar panel. The supply from the cold water cistern to the normal hot water cylinder passes through the solar cylinder where it is *pre-heated* before being further heated in the main hot water cylinder if necessary. In the summer, the solar system

should provide all the heat necessary; at other times of the year, some 'topping-up' will be necessary and, in the winter, precautions have to be taken not only to prevent the solar part of the system freezing (anti-freeze is usually added to the water in this part of the system) but also to prevent the whole system working in reverse which would mean that the domestic hot water cylinder was heating the solar collector!

The solar collector

In order for a solar panel (or solar *collector*) to work efficiently, it must be well insulated which means putting insulation material behind it and a glass cover in front. The panel itself usually consists of a number of metal pipes bonded to a metal plate, the whole of the collector surface being painted matt black.

The glass surface needs to be kept clean – a regular hose-down is all that is required (not always easy on the roof) – and the collector box should be weatherproof.

Controlling solar heating

The main controls fitted to a solar water heating system are temperature sensors, fitted to the water outlet from the solar panel and to the pre-heat cylinder. The sensors control the pump – allowing water to be pumped around the system only when the temperature at the solar panel is higher than the temperature in the cylinder.

To prevent back-circulation, a non-return valve should be fitted into the pipe connected to the bottom of the solar panel.

Installing solar heating

As you can see, much of the solar heating involves conventional plumbing. The main difficulty will be in installing the solar collector itself. The best position for this is at an angle of 45° facing south, but this will be dictated by the pitch of your roof and if the ridge runs north/south, site the collector on the east side to get the benefit of the early morning sun.

Unvented hot water systems

One of the reasons why Continental and American water fittings such as taps and showers cannot be used in this country is that they are designed for use with high-pressure hot and cold water supplies.

In this country, water supplies within a house are at low pressure with the water being supplied from the cold water storage cistern. Hot water storage systems are 'vented' to allow any steam to escape if the system overheats and the cistern accommodates expansion of the water as it is heated.

When new water bye-laws (based on the revised model water bye-laws) came into force in 1989, 'unvented' hot water systems were allowed here with the result that many more foreign fittings will become available. Unvented hot water systems are now covered by the requirements of the Building Regulations as well as the water bye-laws.

How the system works

In an unvented system, *all* the household's water – hot and cold – is supplied directly from the mains. So there is no main cold water cistern in the loft (though a feed-and-expansion cistern may be needed for a central heating system).

To allow the hot water to expand as it heats up, an expansion vessel is fitted (together with an expansion relief valve in case the vessel should fail), and to provide protection against overheating three levels of safety must be built in – thermostat, energy cut-out and temperature/pressure relief valve. The drawing shows the main components:

Line strainer to remove any dirt or other particles which could affect the operation of any of the valves in the system

Pressure-reducing valve to ensure that the actual pressure in the system does not exceed a safe level (typically ⅓ of the test pressure of the hot water cylinder)

Non-return valve (check valve) to prevent hot water returning to mix with the cold water in the mains

Expansion vessel to provide room for the water in the system to expand as it heats up

Expansion relief valve – to allow expanding water to escape in the event of failure of the expansion vessel. Connected to the drains via a *tundish* and must discharge in a safe and visible place

Hot water cylinder must be a special design (steel or copper) to cope with the pressure and normally comes with all the necessary safety devices

Thermostat is the first level of safety. Set to operate at around 60°C to 65°C

Energy cut-out is the second level of safety and is set to operate at 85°C to 90°C to turn off the boiler or other source of heat

Temperature/pressure relief valve is the third level of safety and will discharge water (through a second *tundish* into a safe and visible place) if the temperature of the water reaches 95°C.

What are the advantages?

Apart from being able to use a wider range of taps and fittings (including 'aerated' taps), the main advantages of an unvented system are:
• no cold cisterns and pipes in the loft to freeze and take up space
• no noise from cistern filling
• drinking water at all cold taps
• high and equal pressure at all outlets (giving decent showers)
• no restriction on the placing of the hot water cylinder.

Installing unvented systems

Despite the fact that unvented hot water cylinders often come with all the components pre-fitted to the hot water cylinder, the Building Regulations require the use of an 'Approved Installer' – which means someone who has been on a proper training course.

More of a worry than the installation is the future maintenance of an unvented system. It is vital that the system is regularly maintained to ensure safe operation – and it would be a disaster if someone were to 'bodge' a repair if, for example, one of the two relief valves had discharged water.

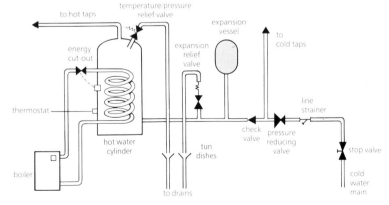

CHAPTER FIVE

HOUSE DRAINAGE

Until now we have been concerned with how water gets into and is distributed around the house, sometimes being heated on the way. In this chapter, it is the turn of the other part of the system: the part which removes dirty water.

This side of domestic plumbing divides into three sections:

Traps which are fitted to basins, sinks, baths and WCs to prevent foul air from the drains entering the house

Pipes to carry the dirty water out of the house

Drains to carry the dirty water underground to the main sewer, septic tank or cesspool.

Unlike fresh water plumbing, which is covered by the water bye-laws, the drainage system is covered by *Building Regulations*, which are administered by the Building Control Officer at your local authority and you must provide details of material alterations or additions that you propose to make to the drainage system of your house.

It is important to recognise that there are three different types of dirty water. First is *waste* water, which comes from washing; second is *soil* water which is the discharge from WCs, and third (the subject of Chapter 6) is *rainwater* landing on the roof or the ground.

Drainage systems

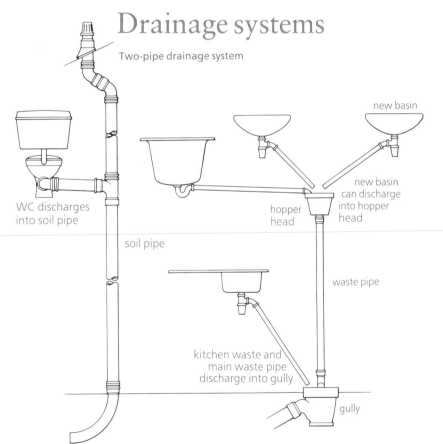

Two-pipe drainage system

WC discharges into soil pipe

new basin

new basin can discharge into hopper head

hopper head

soil pipe

waste pipe

kitchen waste and main waste pipe discharge into gully

gully

At one time, **two-pipe** house drainage systems were designed to keep waste water and soil water completely separate until they entered the underground drains and to prevent 'drain air' entering the house: in Victorian times this was thought to be responsible for all manner of ills. Modern **single-stack** drainage systems allow them to mix above ground, though there are stringent design requirements to ensure that the drains do not pose any risk to health.

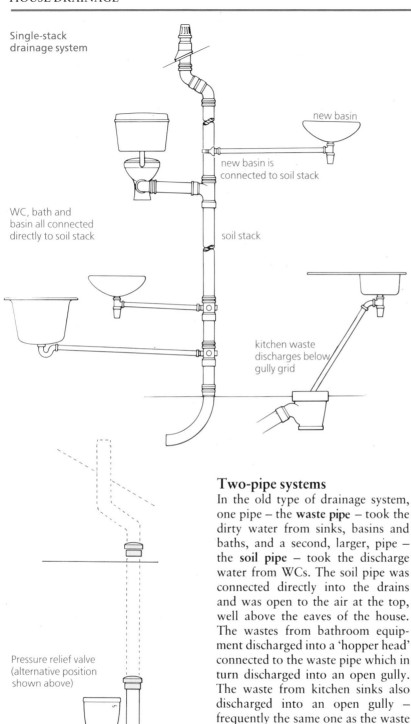

Single-stack
drainage system

new basin

new basin is
connected to soil stack

WC, bath and
basin all connected
directly to soil stack

soil stack

kitchen waste
discharges below
gully grid

Pressure relief valve
(alternative position
shown above)

with dried soap) and the smell from them could enter the bathroom through an open window.

Although this type of system was replaced by the single-stack system around 1960, there are still many houses with the old arrangement.

Single-stack systems

In modern systems, there is only one pipe, to which all the upstairs wastes are connected. This pipe, called the discharge pipe, soil pipe or **soil stack**, runs vertically down to connect directly to the drains; the top of it is open to the air at least 900mm above the top of any opening windows which are within 3m of the stack, unless a *relief valve* is fitted to the top of the stack (see drawing). Originally this type of system had the soil stack running down the outside of the house (as with two-pipe systems); in many houses built since 1976, the soil stack has been placed *inside* the house structure.

Downstairs wastes (from the kitchen sink, for example) can be taken out to a gully as in the older system, except that they must discharge *below* the level of the grid on the top of the gully, which means using a back-inlet gully or one with a hole cut in the grid.

Soil pipes from downstairs WCs may be connected to the main soil stack of a single-stack system, but it is often better to connect them directly to the underground drains.

Single-stack systems are much neater than two-pipe systems and use less pipework. On the other hand, they require careful design to make them work properly and, for this reason, there are strict rules about where and how connections can be made and the size and type of fittings which may be used. They are easiest to design where all the basins, baths, sinks and WCs in a house are close together and/or directly above one another so that pipe runs to the stack can be kept short.

Two-pipe systems

In the old type of drainage system, one pipe – the **waste pipe** – took the dirty water from sinks, basins and baths, and a second, larger, pipe – the **soil pipe** – took the discharge water from WCs. The soil pipe was connected directly into the drains and was open to the air at the top, well above the eaves of the house. The wastes from bathroom equipment discharged into a 'hopper head' connected to the waste pipe which in turn discharged into an open gully. The waste from kitchen sinks also discharged into an open gully – frequently the same one as the waste pipe. Downstairs WCs were connected directly to the drains.

Hopper heads are generally considered unsanitary as they are not self-cleaning (and can get coated

Traps

The traps which are connected to all sinks, basins, baths and WCs (and, incidentally, all wastes from equipment such as washing machines and dishwashers) serve the very important purpose of preventing smells and bacteria from the drains getting into the house. They also stop insects crawling up the waste pipes.

A trap is a specially-shaped piece of pipe in which water remains to make a seal after the basin, sink or whatever has emptied. With WCs, the trap is an integral part of the pan (see page 103); with basins etc., it is connected separately.

The traditional trap fitted to basins was the 'U'-bend which was created by bending lead pipe coming out of the plug hole into a U shape before taking it out through the wall. A screwed access plug or 'eye' in the bottom of the U made it possible to empty the trap of debris if it got blocked. There were also bottle-type traps made from brass, often chromium-plated.

Nowadays, the U-bend type of trap is known as a **P-trap** and is invariably made from plastic, as are all the components of a modern waste system.

Lead U-bend with access plug

S-trap

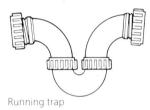
Running trap

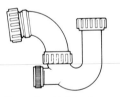
Two-piece bath trap
with screwed connector

There are many designs of trap as a look at any manufacturer's catalogue will show. As well as the P-trap, the three other common types are: the **S-trap**, which has an extra bend in it for when the final outlet is vertical rather than horizontal; the **running trap**, a P-trap with horizontal inlet and outlet; and the **bottle trap** which has a top entry and side outlet with a baffle inside to keep the bottle full. Plastic P-traps and S-traps come apart for cleaning: the bottom of a bottle trap unscrews. Some versions of traps have a capped connector on the side, for connecting an overflow assembly, which can be unscrewed for cleaning. This is the best type of trap to fit where space is limited – with the floor immediately underneath a bath trap, say.

Note that it is possible to use one trap for more than one waste outlet (with a two-bowl sink, for example). The trap must be fitted to the outlet closest to the drains.

In general, P-traps and S-traps are the best type to fit; bottle traps, although compact and neat, can't always cope with large water flows and block more easily. They should not be used with waste disposers.

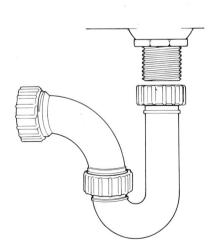
P-trap

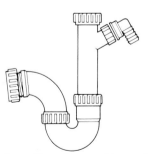
Trap with washing machine waste connector – see page 95

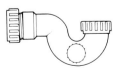
Shallow trap

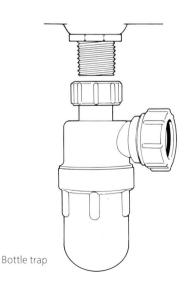
Bottle trap

Depth of seal

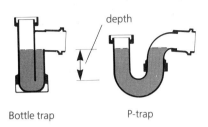

Bottle trap   P-trap

Traps come in different sizes for different diameters of waste pipe, and also in shallow and deep-seal versions. The depth of seal is the vertical distance between the normal water level in the trap and the level at which air can get through the trap (see drawing). For most domestic applications, a 75mm (3in) seal is required; shallow traps (38mm to 50mm or 1½in to 2in) should be used only on two-pipe systems where there is limited space – under a bath or shower tray, for instance. With the possible exception of shower trays, deep-seal traps should *always* be used on single-stack systems: if necessary, the floor will have to be cut away to accommodate them.

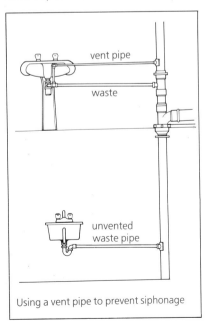

Using a vent pipe to prevent siphonage

It is very important that the water remains in traps – and, in fact, the Building Regulations are contravened if it doesn't. Apart from evaporation (unlikely except in an unoccupied house in hot weather), the main problem is that water can be 'siphoned' or sucked out of traps if the system has not been designed properly to avoid pipes running 'full' and thus causing suction at the trap. This is more often a problem with 'funnel'-shaped basins rather than flat-bottomed baths and sinks and will be worse when there is a vertical drop of branch pipe between the trap and the stack. One way of preventing siphonage is to fit a vent pipe – a separate pipe, connected after the trap, leading to the stack – or to have an anti-siphon valve connected into the wastepipe (see drawing).

There are also *resealing* or 'anti-siphon' traps which may have to be fitted on extra-long waste pipes connected to a single-stack drainage system.

Gullies need to have traps, too, thus providing a second seal between the underground drains and the house. As with WCs, the trap in a gully is built in.

Fitting a new trap

If for any reason the existing trap on a basin, sink or bath needs replacing, it is best to replace the waste pipe at the same time if it is lead, although traps are available which can be connected to materials other than plastic.

The trap is simply screwed on to the bottom of the plug hole fitting (waste outlet), which will have a 1¼in or 1½in BSP screwed thread.

Most traps come with 'universal' fittings which will take different materials and makes of waste pipe. There are two common types of fitting: one where the waste pipe is simply pushed into the outlet of the trap; the second where a plastic 'compression' joint – using a rubber

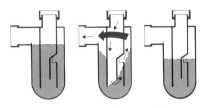

Anti-siphon trap, centre drawing shows what happens when trap is subject to siphonage from the waste pipe

or plastic washer to make the seal – is incorporated into the outlet of the trap.

Some traps have inlets which are adjustable in length and others have special washing machine connections, though a washing machine waste is usually fitted into a trapped stand pipe (see page 95). Kits are sold these days which contain the trap and also the lengths of pipes you need for fitting wastes to basins, baths, sinks, washing machines, etc.

It is important that the waste pipe falls away from the outlet of the waste. The outlet of a P-trap or bottle trap is angled slightly downwards to allow for this.

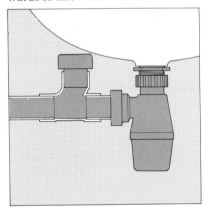

An anti-siphon valve (here one made by Hunter) can be solvent-welded or push-fitted into a ring-seal socket on the pipe to be ventilated. The valve, which must be fitted vertically, comes in two sizes and the internal diaphragm will prevent loss of water from the trap seal by self-siphonage (from the pipe running full) or induced siphonage (from other pipes running full and creating a vacuum)

Waste and soil pipes

The branch pipes leading from individual pieces of equipment, and the main soil stack, are at the heart of the drainage system.

Making alterations and additions to the waste side of a two-stack system is relatively simple: the new branch waste pipe simply needs to be taken out through the wall and angled down into the nearest hopper head or gully. Connections to the soil pipe are more difficult as it will almost certainly be made of cast iron – see page 66.

Before describing the ways in which pipes can and cannot be joined to soil stacks, it is necessary to have a look at the different materials used in waste plumbing.

Materials

The branch waste pipes leading from basins, baths and sinks were traditionally made from lead though copper was sometimes used. These have both been replaced by plastic: polypropylene, which uses push-fit connections, and ABS (acrylonitrile butadiene styrene), which is joined by solvent welding. These are usually available in white or grey in three sizes: 32mm (1¼in), 40mm (1½in) and 50mm (2in); you may also find polypropylene in brown or black. 22mm (¾in) polypropylene is also available and used for overflow pipes for cold water and WC cisterns.

The traditional material for the main soil pipes in drainage systems was cast iron, which is not only heavy but likely to fracture. Hopper heads were often ornate affairs, with perhaps a crest or other decoration on them – this type may still be found second-hand. It is still possible to buy new cast iron fittings for wastes: some have special couplings utilising a rubber gasket to join two lengths of pipe together which does away with the need for traditional sockets and spigot connections,

where one pipe fits into the end of another with caulking compound to make the seal.

Cast iron pipes have now been replaced by uPVC, joined by push-fit connectors or solvent-welding. The standard size for soil stacks is 110mm which is equivalent to the old 4in cast iron waste.

Joining waste pipes

There is a range of waste pipe fittings similar to those used for fresh water plumbing, except that elbows and tees are often slightly more than a right angle (91½° to 92½°) to make sure that the pipes fall properly.

You can also get *flexible* corrugated waste pipe, which is useful for connecting between a trap outlet and the pipe passing through the wall. And you may find adjustable soil and drain elbows, useful for awkward angles.

Solvent-weld joints are made in a similar way to that described for plastic (uPVC) water pipe in Chapter 2. Where the alignment of pipe is important, the pipes and fitting should be assembled 'dry' and pencil marks made. For illustrations of this and other methods, see page 67.

Note that some push-fit joints must be fitted into a solvent-weld system to allow for expansion.

Push-fit or *ring seal* joints are simple to make. The pipe is cut to length and chamfered at about 45° (so that it will enter the fitting more easily) and lubricated with special silicone lubricant before being pushed into the fitting. A pencil mark should be

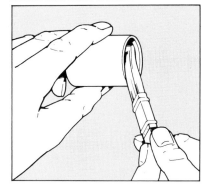

Solvent-weld waste connector

made on the pipe around the mouth of the fitting after the pipe has been pushed fully home so that the pipe can then be withdrawn about 10mm from the fitting on installation. This allows for expansion in the pipe which can be significant when hot water runs through the waste or when the pipe is in direct sunlight.

Compression joints are not the same as those used for copper pipe though they work on a similar principle. Here the sealing 'olive' is made from rubber or plastic; the 'nut' should be tightened up by hand, *not* with any kind of spanner. This type of joint does not allow for any expansion and a ring seal joint should be fitted every 1.8m (6ft) to allow for this.

Waste pipes should be well supported – every 500mm for horizontal runs of 32mm pipe and every 1.2m for vertical ones. Special clips are available.

Note that there is not total standardisation between manufacturers and not all pipes and fittings are compatible. In practice, this means it is best to stick to one brand.

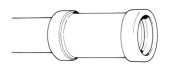

Push-fit waste connector

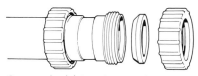

Compression joint waste connector

Connections to the soil stack

The drawing shows the main design points for a single-stack system and the connections which may be made. Detailed requirements are given in BS5572: Code of Practice for Sanitary Pipework.

1 The vent cowl or griddle must be at least 900 mm (3ft) above the top of any opening window (including dormer windows in the roof) which are within 3m of the stack

2 The stack itself should be straight for its main length and offsets are allowed only above the topmost connection

3 The waste from basins will normally be 32mm up to a maximum length of 1.7m (5ft 6in) with a slope of 1¼° to 1½°. The slope depends on the length (see opposite). A 700mm length, for example, can be fitted at a slope of up to 5°. For lengths longer than 1.7m, use 40mm pipe up to a maximum of 3m. For very long lengths use additional venting or a self-sealing trap

4 Pipe supports should be fitted at the correct intervals – usually 1.8m for 110mm pipe

5 WC connections should be swept in the direction of flow with a minimum radius of sweep of 50mm (to the *inside* of the bend)

6 No other connection should normally be made within 200mm of the WC connection. For bath wastes, use an offset connection or a special collar boss on soil manifold

7 Bath and shower wastes should be 40mm with a slope of 1° to 5° for lengths up to 3m. For lengths over 3m and up to 4m, use 50mm pipe. The waste from one pipe should not be able to flow into another

8 Avoid using combined wastes wherever possible: each fitting should preferably be connected separately to the stack

9 Unless taken separately to a gully, sink wastes are the same size as bath/shower wastes. If a waste disposal unit is fitted to a kitchen sink, it should have its own 40mm waste, taken directly to the soil stack with a minimum slope of 7½° and as short a run as possible. Do not use bottle traps for waste disposers

10 The lowest connection to the stack should be at least 450mm above the bottom of the drain. Ground-floor wastes can be taken into back-inlet gullies and ground-floor soil pipes directly into the drains

11 The bend at the bottom of the pipe (underground) should have a radius of at least 200mm. This prevents the bottom of the stack blocking up.

Connections to a single stack drainage system (see text)

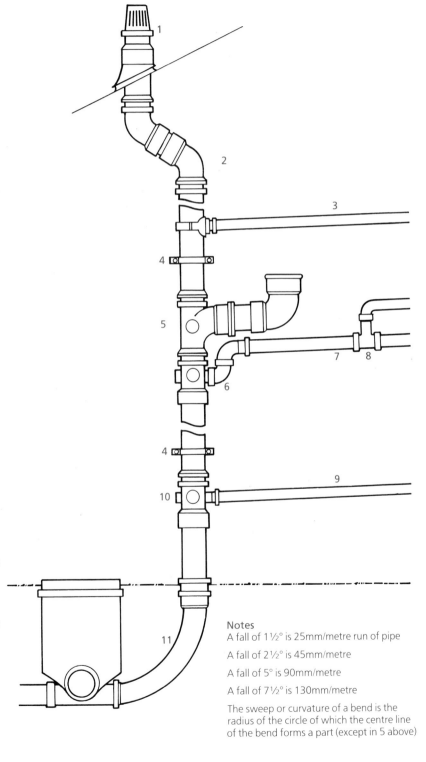

Notes

A fall of 1½° is 25mm/metre run of pipe

A fall of 2½° is 45mm/metre

A fall of 5° is 90mm/metre

A fall of 7½° is 130mm/metre

The sweep or curvature of a bend is the radius of the circle of which the centre line of the bend forms a part (except in 5 above)

Connectors and collars

In order to connect 32mm, 40mm or 50mm waste pipe to an existing uPVC soil stack, a hole must be cut and a boss connected to it. This can be a 'strap-on' boss or a self-locking boss, which is solvent welded to the stack. With either type, you may find a 50mm socket, into which a suitably-sized boss connector or adaptor is welded to take a solvent-welded or push-fit pipe, or a push-fit or compression fitting which takes a branch pipe directly so needs to be the correct size. If there is an existing fitting with a spare 50mm boss socket, this can be drilled out and fitted with a boss connector of the appropriate size.

The new hole can be made with a holesaw or by drilling several holes, removing the disc produced and filing the hole smooth. If necessary, use a padsaw to free the disc, but be careful not to lose it down the stack.

When fitting wastes to a new stack, a bossed pipe or branch is incorporated into the soil stock. These come with single or multiple (usually 2 or 4) boss sockets which will be 50mm size to take boss connectors as above. Alternatively, some sockets may be 32mm or 40mm to take branch pipes directly.

There are special collar bosses or **soil manifolds** which enable a WC connection and a bath connection to

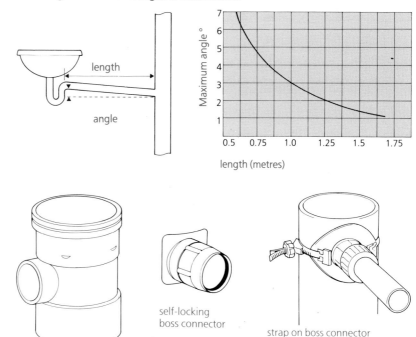

Calculating the maximum length of basin wastes

length

angle

length (metres)

connector bossed pipe

self-locking boss connector

strap on boss connector

be made at the same point on the soil stack. This saves the sometimes awkward task of fitting an offset in the bath waste (which may involve running the pipe below the floorboards) to ensure that it is not connected within 200mm of the WC soil pipe.

Soil manifolds fit into the soil stack underneath the WC swept con-

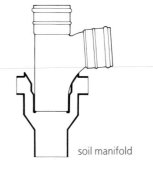

soil manifold

Drain layout

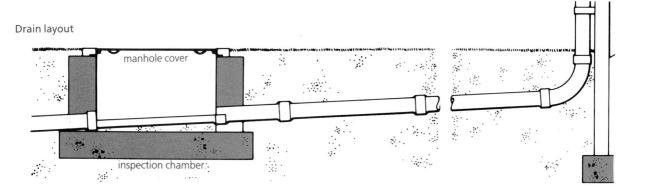

manhole cover

inspection chamber

nection and have an annular (ring-shaped) cavity into which the bath water discharges. The water enters the stack from the bottom of the cavity, merging with the WC discharge which is thus prevented from coming into contact with the smaller diameter bath waste entry.

It is very important that soil and waste pipes have expansion joints between fixed points: check the manufacturer's instructions.

Warning: when working on the soil stack, make sure that no-one inside the house empties a basin or flushes the WC.

Connecting into cast iron stacks

This is a much more difficult job than connecting into uPVC stacks – and is possibly a job that you would want to leave to a plumber.

Joining into a cast iron soil pipe will mean cutting a length out of the stack (with a hacksaw or angle grinder) to fit a new length of pipe with a branch connection. Modern sleeved couplings avoid the need to dismantle the whole stack.

Connections must follow the rules shown on page 64. Often there will not be room to connect a downstairs WC soil pipe with the correct fall and a new soil pipe will be needed running directly to the drains.

You will probably have to do a bit of hunting around in order to find the fittings for making connections to cast-iron stacks: if local merchants don't have what you need, contact a specialist firm.

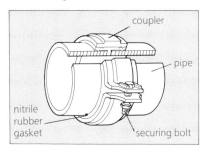

Cast iron coupler
(Glynwed Timesaver)

Two ways of connecting WC and bath wastes

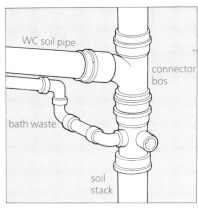

Using conventional boss

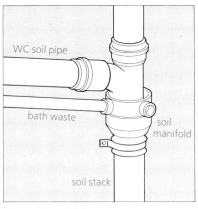

Using soil manifold

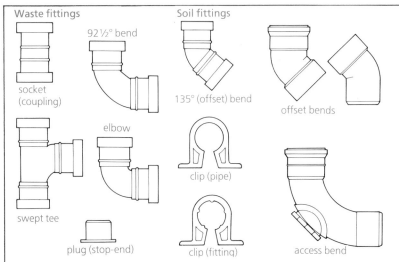

Waste and soil fittings

The manufacturers of waste and soil fittings produce excellent catalogues with details of their various fittings and give advice and information on jointing methods and ways to design the system. There are many different types of fitting. The main ones are:

Sockets for joining two lengths of pipe together: both push-fit and solvent-weld available. Some types have one plain end to take a second connector

Elbows for taking pipes around corners. Angles of 90° and 135° are available as well as adjustable and access elbows. Also 91¼° and 92½° bends for joining horizontal pipes to vertical pipes to give a fall to the horizontal length

Tees available mainly in 92½° and 135° Tees are usually **swept** and should be fitted so that the sweep goes with the direction of flow from the branch into the main pipe. Some branch pipe connectors (for soil pipes) have removable access covers for cleaning

Connectors are available in various types including *reducers* for connecting different diameters of pipe, and fittings for connecting to copper pipe and screwed outlets

WC connectors come in various patterns. The **Multikwik** connector (see Chapter 7) provides the most versatile method of connecting a WC pan to the soil branch pipe.

Clips for all sizes of pipe

Joining waste pipe

Solvent-welding

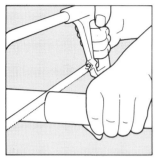

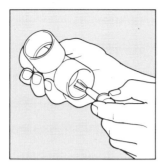

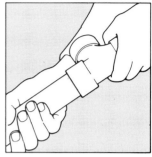

Cut the pipe square with a fine-toothed saw, using paper as a guide, and remove all rough edges with fine glass paper

Push the pipe into the fitting and mark the amount that is inside with a pencil. Clean the pipe surface with solvent

Apply solvent cement to the surface of the pipe and to the inside of the fitting (which should also be clean)

Push the pipe into the fitting, twisting it slightly, and clean off excess cement. Leave for 24 hours before use

Push-fit (ring-seal) joint

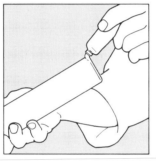

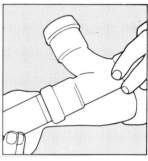

Cut the pipe square as above and form a 45° chamfer on the end of the pipe using a file or special tool. The chamfer should be even

Remove all burrs of plastic and smooth the ends of the pipe with glass paper and apply silicone lubricant to the pipe

Check that the rubber seal in the fitting is in place and secure in its seating, ensuring that the groove is clean

Push the pipe into the fitting squarely until it goes no further. Mark the pipe with a pencil and withdraw it about 10mm

Compression joint

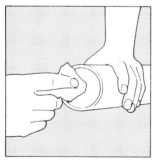

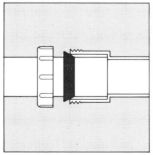

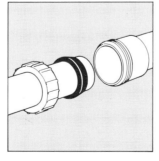

Cut the pipe as before and smooth the ends. The end of the pipe does not need to be chamfered for this type of joint

Most brands of compression waste fitting will accommodate different types of pipe with slightly different diameters

Unscrew the locking ring and position on the pipe followed by the sealing ring (make sure this is the correct way round)

Apply liquid soap to the pipe and push it into the fitting. Tighten the nut up by hand – do *not* use a spanner

Connecting to the soil stack

Connecting to the pipe

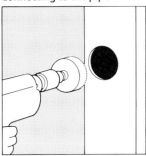

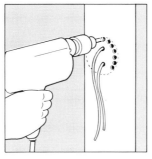

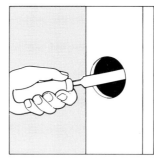

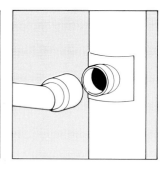

To connect into an existing stack a hole must be drilled (*not* within 200mm of the WC connection). A hole saw can be used for this

Alternatively, drill a circle of holes and remove the disc made. The two centre holes allow a retaining wire to be fitted

Smooth off the hole and fit the appropriate strap-on or self-locking boss. Most bosses are solvent-welded in place

The pipe can now be connected into the boss, with a suitable connector if required by solvent-welding or with a ring seal joint

Connecting to a boss shoulder

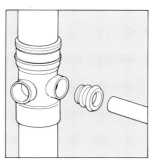

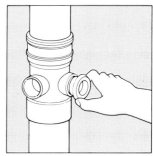

Where there is an existing unused boss socket, this can be drilled out using a hole saw

The pipe is connected using a boss connector solvent-welded in place with the pipe a push-fit or solvent-weld in the connector

Remove the rough edges from the hole and use solvent cleaner before applying solvent-weld cement to both surfaces

Push the connector firmly into the boss shoulder making sure it is aligned properly and remove surplus cement

Taking pipe round corners

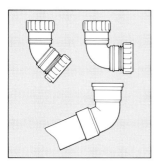

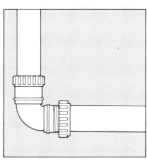

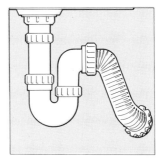

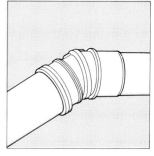

Two of the different angles of bend that are available: 90° and 135°. Also shown is an *offset* elbow

A 92½° (or 91½°) bend ensures that a horizontal pipe falls away where it is connected to a vertical section of pipe

Flexible waste pipe is useful for making the connection between a trap and the pipe passing through the wall

Adjustable soil pipe connectors can be varied to give an angle of up to 25°. Useful for connecting WC waste pipes

The drains

Typical single house drainage

Communal drains

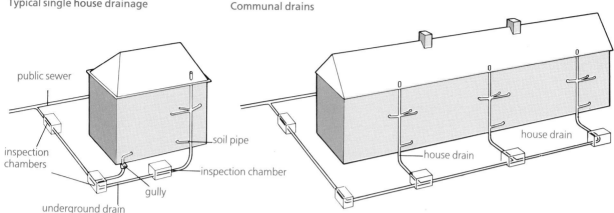

Where drains are connected to the public sewer, they can either be *single*, where each house has its own connection, or *communal,* where each house is joined to one big pipe which goes to the sewer. Communal drains save time and money when several houses are being built together, but there can be a problem sorting out whose responsibility it is when something goes wrong.

Gullies

All gullies, whether they take waste water from upstairs fittings (via a hopper head) or waste from kitchen sinks, must have traps to prevent smells coming up from the underground drains. In modern systems, pipes leading into gullies must be below the level of the grid but *above* the level of water in the trap. This is to ensure that the gully does not get blocked by leaves and other debris and also so that the action of the water gushing into the gully is to clean it. This can be achieved by passing the pipe through a hole cut in the grid, but a special type of gully, which includes the waste pipe connection as part of the gully, will also meet this requirement. These are known as **back-inlet gullies**. Existing

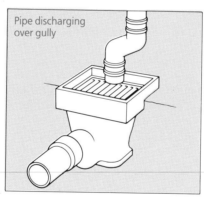

Pipe discharging over gully

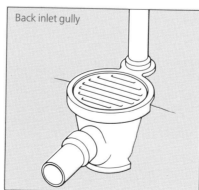

Back inlet gully

gullies can be modified by fitting a wire mesh over them with a hole cut to take the waste pipe.

From the gully a pipe leads to an **inspection chamber** which may also have the soil stack connected to it.

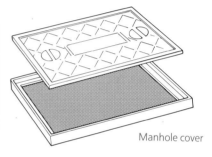

Manhole cover

Inspection chambers

With underground drains, it is very important that access should be available in case the drains ever get blocked. This is usually achieved by having an inspection chamber – fitted with a **manhole cover** – wherever a connection is made to the drain, wherever there is a junction between drains or wherever the drain itself changes direction or gradient. Drains should ideally be laid in straight lines between inspection chambers and should always fall away from the house with the correct gradient to ensure that the flow through the pipe is fast enough to ensure solid matter is carried along. Recommended minimum gradients for drains are 1 in 80 or 1 in 40, depending on the likely peak flow rate.

The layout of the drains can usually be discovered by pouring water down various sinks, WCs, etc and seeing which inspection chamber it comes into. The local authority may also have a plan of your drains.

The inspection chamber is normally rectangular with cement-rendered brick-built sides. Originally, rendering was on the inside, but this can flake off and block the drain, so rendering (if used) is now put on the outside. At the base of the chamber is a **half channel**, often consisting of half a piece of drain pipe with half channels coming in from branch drains. Between the channels, all of which will be 'swept' in the direction of flow, the chamber is built up with concrete **benching** so that any splashes are directed back into the channels. The top of the inspection chamber has a metal frame into which the manhole cover fits. This usually just rests in place (it's rather heavy), but where inspection chambers are inside a house (unusual except where an extension has been built on), the manhole is screwed down and has to be specially sealed.

In modern drainage systems, an inspection chamber may not be necessary at all junctions and bends: a **rodding point** may be provided instead. This is a length of pipe, with a gentle bend in it, which leads up to the surface where it is fitted with a removable cover. If the drains get blocked, this cover can be removed and drain rods pushed down the rodding point to clear them.

In older systems, the inspection chamber nearest the boundary of the property may be fitted with an **interceptor trap** through which the water will normally flow: a rodding arm is fitted with a stopper which can be used to unblock this length of drain. The purpose of this trap was to keep sewer gases (and rats) out of the household drains, but modern drain systems don't need this extra trap. This type of inspection chamber may also be fitted with a **ventilator** with a hinged mica flap – the idea of which was to let fresh air into the drains but not let foul air out. In practice, many got blocked or damaged and some will have been removed – though their use is still mandatory in some areas. Your local authority will be able to advise you on the requirements for fresh-air ventilators.

Drain materials

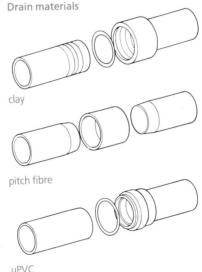

clay

pitch fibre

uPVC

Drain materials

Old drains systems were constructed from 2ft lengths of glazed clay pipe. Originally these were jointed with a gasket of tarred hemp between the 'spigot' (male end) of one pipe and the 'socket' (female end) of the next one. Although providing a degree of flexibility to cope with ground settlement, this type of joint tended to leak and a cement filling was added to it with the whole pipe laid on a bed of concrete carried up the sides. This method had the problem that ground movements could cause fracture of the fragile pipe. Modern drain design utilises materials and joints which are more flexible, laid on a non-rigid 'bed'.

Clay pipes are still used but they are not usually glazed. Joints tend to be of the rubber 'O'-ring type fitted between the spigot on the end of one length of pipe and the socket on the next. Plain-ended pipes are joined with a plastic sleeve which slips over both ends of pipe and uses rubber sealing rings to make the joint. The main problem with this type of pipe is that it is difficult to cut to length accurately. Although it is made in relatively short lengths, the traditional way to cut clay pipe is with a hammer and chisel, which takes a bit of practice. Filling the pipe with sand will help to prevent it breaking in the wrong place. If you own one (or can hire one), an angle grinder would make a neater job.

Concrete inspection chamber

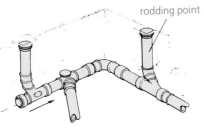

benching

half-channel

Rodding points at bend and junction

rodding point

Inspection chamber with interceptor trap

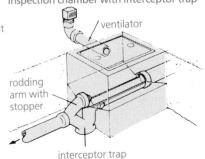

ventilator

rodding arm with stopper

interceptor trap

Pitch fibre pipes are made from waste paper and other fibres soaked in pitch. When used for drains, this type of pipe usually has a plain end and is jointed with a plastic sleeve containing sealing rings. These are often known as 'snap' rings as they will suddenly and obviously snap into place as the pipe is pushed into the sleeve. This type of joint will allow a certain degree of misalignment when the pipe is laid. Pitch fibre pipes are no longer used for new installations.

Plastic pipes are the most common these days in new underground drainage systems. The uPVC pipes can be joined in the same way as uPVC waste and soil pipes – that is by solvent-welding or by ring seal (push-fit) joints – but it's better to use push-fit joints for flexibility. A range of fittings is available including ready-made inspection chamber bases, rodding points, gullies and fittings to join uPVC to other materials. Pipes in uPVC come in long lengths (typically 3m or 6m) but can be cut very easily.

The standard size for plastic drain piping is 110mm, though larger sizes (160mm, for example) are also widely available. Smaller sizes of drain pipe (82mm, for example) can be used where the drain will carry only waste (and not soil) water.

Laying new drains

Whether you are replacing an existing faulty drain or laying a new one, perhaps to lead from an extension, the major part of the job will be excavating the trench. Since the work is covered by Building Regulations, you will need to inform your local authority; they may well tell you the type of trench necessary, the ways the drains have to be laid in them depending on local soil conditions and the material to be used for the drains. As with any type of building work requiring Building

Polypropylene inspection chamber

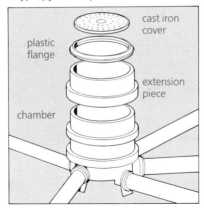

Regulations approval, the Building Control Officer will want to inspect the work at various stages as well as approving the plans. He or she will almost certainly want to test the drain to make sure it is watertight when the job is finished.

The maximum depth of the trench will be at the point where it enters the main sewer; otherwise the trenches will need to be deep enough to allow for a fall of the required gradient away from the house. If this is not possible because the drain is too long or the land falls the wrong way, the pipe used for the drain will have to be larger and laid with a gentler slope. The trench must be about 500mm wide with 100mm added to the depth for any bedding material which the Building Control Officer may insist on. The usual bedding material is coarse aggregate (gravel).

Unless you have some experience in cutting clay pipes (or can arrange the layout so that no cutting is necessary), it is probably best to stick with the easier-to-use uPVC. The manufacturers' catalogues provide information on the range of fittings (gullies and so on) that are available and should provide instructions on how to join and lay pipes.

To join a new length of pipe into an existing inspection chamber, you will have to break a hole in the wall

of the chamber and chop away at the benching so that a new half channel can be laid to join up with the main channel. The benching will need to be made good with new mortar (benching is designed to have a slope of 1 in 6) and the walls of the inspection chamber repaired with small pieces of brick mortared in place. The mortar mix should be between 1:1 and 1:3 (cement:sand).

Putting in a new inspection chamber will involve digging quite a large hole in the ground – and supporting the sides of it while the chamber is being built to prevent them caving in. The brick sides of the chamber and the concrete benching should all be laid on a firm bed of concrete at least 100mm thick. If the new inspection chamber is to be positioned over an existing clay drain, the difficulty is likely to be cutting into the existing drain.

The job of installing a new inspection chamber may be made easier if a pre-cast concrete inspection chamber or one of the modern glass-reinforced plastic inspection chambers is used. Check with your local authority that these are allowed.

When a Building Control Officer tests a drain, he or she will carry out either an *air test* or a *water test*. In an air test, stoppers are used to block off the ends of the drain and air is blown into the drain to see whether it can hold the pressure. In a water test, the drain is also stoppered and then filled with water to see whether this can be retained. You could carry out your own water test.

The drain should not be covered over by backfilling the trench until the test has been completed.

Backfilling consists of putting in a layer of aggregate followed by layers of earth, tamping down each layer before shovelling on the next. Some settlement is likely to take place over a period after the trench has been filled, but this can be filled in at a later stage.

Building over drains

When planning a new extension to your house, it is more than likely that the best position for it will be right on top of the existing drains manhole covers. Once again, the most important person to talk to will be the local Building Control Officer. He or she may say that the existing inspection chamber should be closed up and a new one installed outside the extension. Alternatively, he or she may suggest that the inspection chamber is left in operation, fitted with an airtight cover bolted or screwed down so that it can be removed if necessary. Special recessed covers are available which allow you to fit a permanent floor covering.

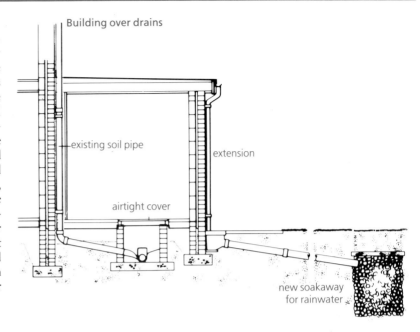

Building over drains

existing soil pipe

extension

airtight cover

new soakaway for rainwater

Sewers, cesspools and septic tanks

The majority of houses are connected to the public sewer and this is certainly the best way to dispose of household waste and soil water. In some rural properties, however, the cost of joining the house to the public sewer may seem prohibitive, but you should balance this against the continuing cost of having a cesspool or septic tank emptied.

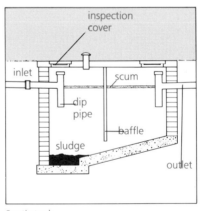

inspection cover

inlet

scum

dip pipe

baffle

sludge

outlet

Septic tank

inspection cover — vent

inlet

Cesspool

Cesspools are simply lined holes in the ground which act as reservoirs for sewage until they are emptied by the local authority.

A 2.25 cu m (500 gallon) cesspool could fill up in as little as a week with a family of four living in the house; a 18 cu m cesspool would last for at least a month and is the minimum size recommended.

Originally, cesspools were built from brick rendered with concrete on the outside. This was followed by construction from a series of concrete rings, 6ft in diameter, mortared together; more recently, cesspools have been made from glass fibre (glass reinforced plastic, GRP).

Septic tanks are more like miniature sewage works and rely on the action of bacteria to break down the sewage into harmless liquid and sludge. Usually, two chambers are needed: the first where the bacteriological action takes place and the second where the resultant liquid is filtered before being dispersed through the subsoil via land drains or disposed of

into a ditch or stream. It is very important that the flow into and out of the septic tank takes place below the level in the tank so that the *scum* on the surface is not disturbed. Special 'dip-pipes' ensure this. Some GRP tanks have three chambers.

A well-constructed septic tank should need little maintenance apart from emptying once a year, but it is important that not too much disinfectant, bleach or detergent is used in the house – otherwise, the bacteriological action could be affected and the tank stop working.

Laying drains

The trench will need to be around 500mm wide and 600 to 700mm deep – allowing for 100mm of bedding material

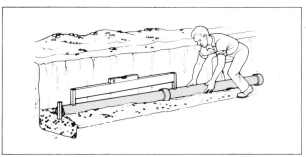

Put down the gravel fill layer (if required) and tamp down. The slope of the gravel should be shallower than required

Position the next pipe and push it into the joint. Use a spirit level and gauge board to lay the pipe at the correct angle

Where the new waste/soil pipe passes through the house walls, fit a lintel over it. Use a long radius bend to the drains

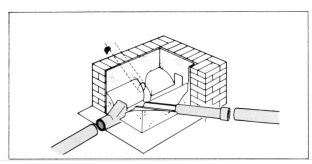

Break a hole into the existing inspection chamber and chop away the benching. Bed new pipe in mortar over the channel

Follow the manufacturer's instructions for making the pipe joints; make good the sides of the chamber and the benching

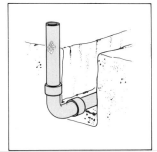

To test the drain, fit a vertical length of pipe plus pipe stoppers and fill with water. Check after 30 minutes for leaks

After the Building Control Officer has tested the drains, the trench can be backfilled. Use gravel (or aggregate) for the first layer

Unblocking sinks and drains

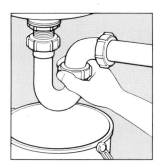

A plastic sink trap is easy to undo by hand. Place a bucket underneath and poke wire about inside to remove blockages

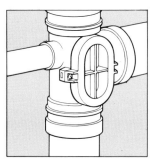

Soil and waste pipes often have access doors which allow lengths of the pipe to be cleared. Stand to one side when opening

If a gully is overflowing, try to unblock the grid using a length of stick to clean it out and rinse the gully clean

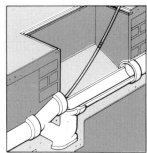

The rodding arm of an interceptor trap needs to be used only when it is the final length of the drains which are blocked

Clearing blocked sinks, WCs and drains

The symptom of a blocked **sink** (or basin or bath) is fairly obvious: the dirty water refuses to empty out. Usually, but not always, this will be the result of a clogged trap or of grease built up on the inside of the waste branch pipe and is relatively easy to deal with. Blocked WCs and drains are a different proposition and care should be taken when working; also make sure you disinfect rubber gloves (essential to use them) and equipment afterwards.

If a **WC** is blocked, the problem could be a blocked soil pipe. These usually have access doors to enable drain rods to be inserted. Be careful when taking the door off.

The first signs of a blocked **drain** are likely to be overflowing gullies or manhole covers or a blocked downstairs WC. To find where the blockage is (unless it is simply the gully itself which is blocked), you will have to lift the inspection chamber manhole covers starting at the house and working outwards. The covers may have rusted badly and the handles broken off, in which case the best tool to lift them is a garden spade inserted under the rim, followed by a stout length of wood. It may be necessary to clean round the cover first. Take the opportunity to clean out the channel the cover fits in and to fill it with grease. Not only will this help to make the cover easier to lift next time, but it will also improve the seal. (Also rinse out all the inspection chambers with a garden hose once the blockage is cleared.)

As you work along the chambers, you should find one that is empty, which will tell you that the blockage is between that one and the previous full one. If they are all full, the blockage is between the last inspection chamber and the main sewer. Check with neighbours that they haven't got the same problem — something could be wrong with the main sewer — before attempting to unblock your own drains.

It may be that there is no actual overflowing, but rather a strong smell of drains — particularly by the front gate. This could quite likely be due to a blockage in the final inspection chamber — particularly if it is fitted with an interceptor trap. The rodding arm plugs for these quite often fall out and block the trap, which means that the chamber will empty to the level of the rodding arm, but the sludge which is left will cause the smell.

Although you should be able to clear most blockages yourself, there are some — a tree root growing into a cracked drain for instance — which require specialist help. Likewise, if you find you will need specialist equipment, even though this is easy to hire, you may prefer to leave the whole job to a specialist firm.

Blocked sinks

The first thing to try with a blocked sink, bath or basin is a **pressure** device. The simplest — and often the most effective — is the sink plunger. This is held over the blocked plug hole and worked up and down several times to force water down the waste pipe. Plungers require 75mm (3in) or so of water in the sink to make them work. There are various hand pumps which work on the same principle. When using a plunger, ensure that it seals well on to the surface (smearing petroleum jelly on to the rubber surface helps) and that you don't simply pump the water up the overflow pipe and back into the sink — hold a wet cloth tightly over the overflow while using one.

If a plunger doesn't work, you may have to remove the trap or the cleaning eye of the trap. On modern waste systems this is relatively easy, but on old-fashioned systems it may be anything from difficult to impossible — and care needs to be taken not to crack a ceramic basin when unscrewing a tight cleaning eye. Some plastic traps and other fittings (such as elbows) have a plug which you can remove to poke around inside.

Put a large bucket underneath the trap to catch the contents of the sink — and remember *not* to pour this away until the trap has been refitted!

If neither of these methods works, it is likely that fat or grease is blocking the branch waste pipe further along. The answer here is to use some kind of **plumber's snake** – a length of flat or helically-wound spring steel with a brass knob or a steel spring on the end which can be inserted into the waste pipe and rotated. The snake is used to move the blockage nearer to the drains until the passage is cleared.

Plumber's snakes can be hired, though might be a tool worth owning — you may be able to improvise something from a length of net curtain wire with a hook at the end. A bent metal coat hanger may work for blockages close to the trap but, unlike plumber's snakes and curtain wire, it will not go round bends.

There are also **chemical** drain clearers. Many of these are based on caustic soda and require care in use and protective clothing, gloves and goggles. They are not as effective as mechanical methods and should *not* be used where the waste is fully blocked: the result could be a basin full of aggressive chemical. Chemical clearers can, however, be used when the main blockage has been partially cleared to give the system a good clean and flush out.

It is a good idea to keep sink and basin wastes clear by occasionally pouring some washing soda down the plug hole and by being careful not to put fat, vegetable matter, tea leaves or rice down the sink.

Blocked WCs

If a WC is blocked, a large sink plunger may work (don't use one with a metal flange which could damage the pan), provided it has a long handle; alternatively, use a length of drain rod fitted with a rubber disc.

Blocked drains

Once you have found the blockage (see opposite), you will need a drain clearing set (this can be hired) consisting of rods which are screwed together with a choice of accessories to fit on the end. These include plungers, wormscrews (for boring into a blockage and pulling it out), and scrapers for removing sludge and silt and either pushing it down the drain or pulling it back into the inspection chamber on which you are working.

You select which device you want, put it on the end of the drain rods and push them down the drain or the rodding point. Finding the hole in a blocked inspection chamber is not always easy: the way to do it is to find the bottom channel and to push the rod along this until it goes down the drain. Finding the rodding arm in an inspection chamber fitted with an interceptor trap is particularly difficult. Not only do you have to find it, but you may have to knock out the retaining plug. For *next* time, attach a piece of wire to the retaining plug and nail it to the sides of the inspection chamber (with a wire staple) near to the top so that you can simply pull it out. It is possible that a piece of slate or glass has been cemented over the rodding arm. Smash this to get the drain rod in.

It may help to twist drain rods when pushing them down the drain. Do this in a *clockwise* direction, *not* anticlockwise or they will unscrew and you could have a length of drain rod stuck down the drain – as well as the original blockage!

You may also come across larger versions of plumber's snakes sold for clearing drains, including electrically-powered ones. These work well, but it is probably better to hire drain clearing equipment rather than buying it – so your choice will depend very much on what the local hire shop has available.

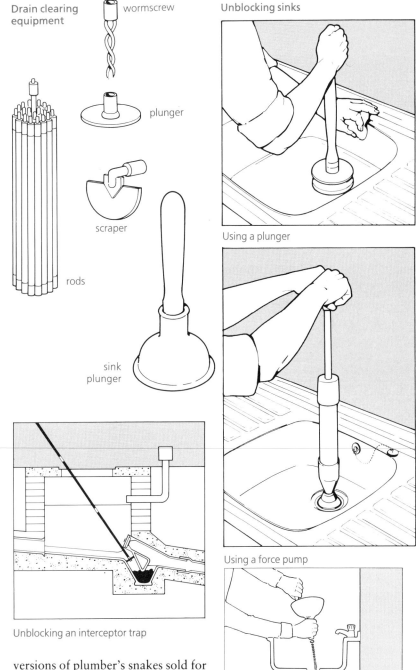

Drain clearing equipment

wormscrew

plunger

scraper

rods

sink plunger

Unblocking an interceptor trap

Unblocking sinks

Using a plunger

Using a force pump

Using a plumber's snake

CHAPTER SIX

RAINWATER DISPOSAL

The arrangements for getting rid of rainwater are often skimped by builders who will put in the cheapest possible system — and not always install it properly. In time this will lead to the development of faults.

Faults can range from the merely inconvenient, when water spills out of sagging or blocked gutters on to unsuspecting people underneath, to the more serious problem of penetrating damp, where peeling decorations and damp patches inside the house can often be traced to a leaking gutter or downpipe outside.

So it's important that you know about your rainwater disposal system — so that you can either put it in good repair and keep it that way or safely replace an old system with a modern plastic one which will work properly.

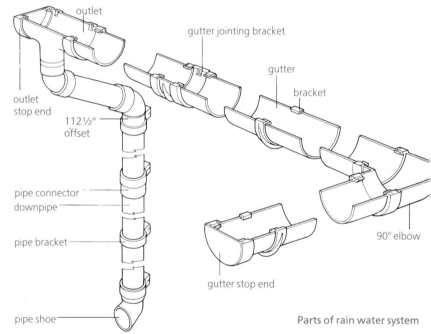

Parts of rain water system

The system explained

After the rain has fallen on the roof, it runs down into **gutters** under the eaves. The ends of any roofing felt overlap into the gutter, and the gutters generally run at a slight slope to the **downpipe** which carries the water away. The individual lengths of gutter are connected with **joints** and are secured with **brackets**. At the end of straight lengths, there will either be a **stop end**, an **elbow** (to take the gutter around a corner) or a

stop end outlet to connect to the downpipe. Mid-run connections to the downpipe are known as **running outlets**. With both types there often needs to be an offset (or **swan neck**), to connect the outlet to the top of the downpipe, so that this can be positioned against the wall.

In older systems, downpipes may lead into hoppers (or **rainwater heads**) which are connected to a single downpipe discharging via a

shoe over a gully. As with waste systems, having the discharge *over* the gully can be a nuisance if the gully grid gets blocked with leaves. All modern systems must have the discharge below the grid.

In some areas, the downpipes lead to **trapped gullies**, which in turn take the water to the inspection chambers connected to the main drain leading to the sewer (see Chapter 5 for details). In many areas, the volume of rainwater would often lead to problems as the main sewer could not cope, so here the downpipes are connected to a **drainshoe**: this is connected directly to a separate drainage system.

The third choice is for each house (or pair of houses) to have its own **soakaway** – this is simply a hole in the ground filled with bricks or rubble, into which the rainwater is taken and out of which it will slowly disperse. Soakaways can cause problems if they have not been designed properly or are not big enough. If you build an extension, a soakaway will often be required.

Materials

Some early rainwater systems were made from lead but there are few of these left. **Cast iron** superseded lead as the main material used for gutters and downpipes but cast iron systems are now likely to be showing signs of leaking at the joints between lengths of guttering or at the joints in down-pipes. The material itself may also have broken or be rusting.

Although it is possible to buy spare parts for cast iron guttering and to repair or replace leaks at joints (and even to repair split or corroded sections), there comes a point when it might be better to replace the whole system.

Another material used was **asbestos cement**. This is similar in size and shape to cast iron systems and jointed in the same way. Galvanised steel guttering has also been used.

The most common material for new houses and as a replacement for cast iron is uPVC – a type of **plastic**. This is considerably cheaper and lighter than cast iron and the different makes (although not always compatible with each other) are all simple to join. Grey and black are the most common colours for plastic guttering, though some brands are available in brown, green or white. You could, of course, paint it any

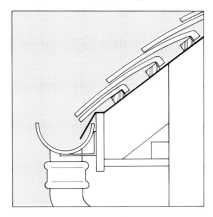

Roofing felt overlapping into gutter

colour you wanted, but painting is not necessary for protection.

Most plastic guttering is smooth, but some brands are ribbed inside, which improves the rigidity and allows some water to flow even when leaves are blocking the gutter. The main disadvantage of plastic as a guttering material is that it can't take the weight of a ladder; but it is risky leaning a ladder even against cast iron so if you need to support the top of the ladder away from a house wall, it is much better to use a ladder *stand-off*.

Another material used for guttering is **aluminium**, which is sometimes joined with silicone or butyl-based sealants and sometimes put up in long lengths by the supplier. Aluminium guttering is available with a choice of coloured coatings.

There are three main shapes for gutters used in rainwater systems: round, square and 'ogee'. Round guttering is shaped like half (or slightly less than half) of a circle and is used with circular downpipes. Square guttering is used with rectangular downpipes. Ogee guttering (common in cast iron systems, but also available in uPVC and aluminium) has an S-shaped front, and is usually fitted with round downpipes. Other moulded shapes are available.

Some manufacturers make 'deep-flow' gutters for large houses.

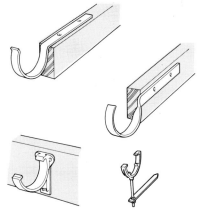

Gutter brackets: *top* rafter brackets for cast iron; *bottom* plastic fascia bracket, rise-and-fall bracket

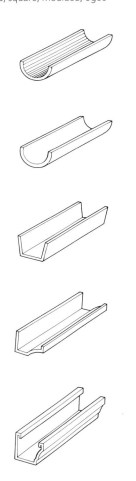

Gutter profiles: *from the top* ribbed, half-round, square, moulded, ogee

Repairing gutters and downpipes

Cast iron systems should be regularly painted (gloss paint outside, bitumastic paint inside) to prevent corrosion, and all gutters should be given a good clean out every autumn to get rid of leaves, dirt, moss and tennis balls. Remove all debris with a trowel and don't push it into the downpipe. A wire 'balloon' should be fitted into the top of each downpipe to prevent birds nesting and to keep out leaves and other debris.

Eventually, though, old cast iron systems will need repair. Repairing gutters is not a job for the fainthearted since it involves working up a ladder for quite a long time. A hired platform tower makes the work safer, but if you have any doubts about your ability to work at heights, it would be better to call in a guttering contractor to do the job for you. Also note that cast iron guttering is very heavy and can cause injury or damage if it falls.

Repairing leaks in gutters

Cast iron gutter lengths have a 'socket' on one end, inside which sits the non-socket (or 'spigot') end of the next length. The joint is made with putty or mastic and a gutter bolt holds the two lengths together. If putty is used (as was common on older systems), this will dry out in time and the joint may leak.

In order to separate the joint, the gutter bolt must first be removed. If it can't be unscrewed (it may well have rusted solid), cut through the bolt with a junior hacksaw flush with the bottom of the gutter.

Next remove all the old putty from both parts of the joint and give the metal surfaces a good rub down with a wire brush and, if necessary, abrasive paper before applying two coats of bitumastic paint. To reassemble the joint, use a non-setting mastic rather than putty and apply it liberally to the socket. Fit a new gutter bolt and tighten it up. This will cause some mastic to squeeze out: scrape this away from both the top and the bottom.

The joints between plastic gutters can leak too – usually because dirt or grit has got between the gutter and the seal or because the seal itself has perished. Separating the two parts is considerably easier than with cast iron. Clean out the joint and, if necessary, replace the seal: you will probably have to take a sample of a joint along to your merchant in order to find a matching replacement. Mastic sealant can also be used to seal gutter joints.

Repairing sagging gutters

If a gutter is sagging and allowing water to spill out, it is likely that one of the supporting brackets has broken or bent. A new bracket can be fitted fairly easily but new galvanised or alloy screws should be used. Make sure the fascia board is sound, and check the joints at the ends of the sagging length – they may well need renewing. Rise-and-fall brackets (see page 77) provide a method of adjustment for sagging gutters.

Repairing broken gutters and downpipes

The best material for repairing splits and cracks in cast iron gutters and downpipes is the kind of resin filler used for repairing cars.

The area around the crack should be thoroughly cleaned and the two-part resin filler applied after it has been mixed up in accordance with the instructions. Slightly overfill so that the surface can be rubbed down smooth once the filler has set. After rubbing down, apply bitumastic paint inside and outside.

Larger holes or splits may need reinforcing with glass fibre bandage.

Clearing blockages

If there is a blockage somewhere in the downpipe, water will spill out of the gutter.

Sometimes the blockage may be visible from the top of the downpipe and can be hoicked out from the top with a length of stiff wire – try a straightened-out coat hanger.

Often, directing the flow of water from a garden hosepipe into the top – or bottom – of a downpipe or an open hopper will clear the blockage; leave the hosepipe running for a while to give the system a good flush out. If all these measures fail, it may be necessary to use a pole with a rag securely tied on to it or clearing rods similar to those used for clearing drains. Remember to put something at the bottom of the downpipe if it discharges over an open gully in order to catch whatever is blocking it, or else you might block the gully or the drains further along.

Repairing leaks in downpipes

If a downpipe is damaged, it may be necessary to remove it from the wall and to replace (or repair) lengths and remake the joints. Getting an old cast iron downpipe off the wall is not easy and there is a fair chance that if it didn't need replacing when you started, it will by the time you've removed it. The downpipe brackets are held in place with **pipe nails** hammered into wooden blocks in the walls, which you will need to lever out. You may also need to apply heat to the pipe in order to separate the joint. Put in new wall plugs and screw the brackets in place.

It may well be worth trying to gum up the joint with a mastic sealant or self-adhesive bitumastic flashing rather than going to all the trouble of dismantling the system: use the sealant on the joints even if you do dismantle and reassemble them.

Repairing leaks in cast iron gutters

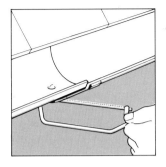  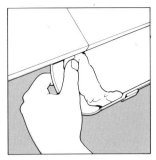 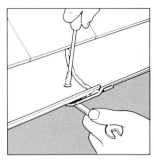

Remove the old gutter bolt. This may have to be sawn off with a junior hacksaw if it has rusted in place

Scrape off all the old putty from inside the gutter and clean both ends with a wire brush. Apply bitumastic paint

Use non-setting mastic to make the new joint. Apply liberally to the inside of the gutter and reposition the second length

Tighten up the gutter bolt using a screwdriver and spanner until it is tight. Remove excess mastic with a putty knife

Repairing cracks in cast iron gutters and downpipes

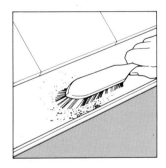   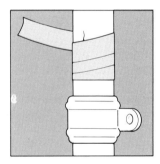

Before starting the repair, remove all debris and loose rust and rub down with a wire brush and abrasive paper

For large cracks in gutters, use glassfibre bandage to reinforce the repair – this is 'buttered' with resin filler from the outside

When the bandage has set, fill the crack with resin filler using a flexible filling knife. Rub down when set, and paint

For downpipes, glassfibre bandage can be wrapped around the split and more resin filler applied if necessary

Maintaining gutters

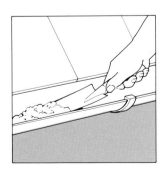  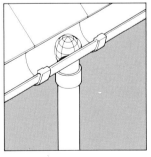 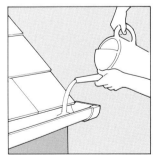

Remove debris from the inside of gutters with a trowel at least once a year. Don't push it down the downpipe

If downpipes get blocked you could try pushing down a pole with a rag tied to the end. Alternatively, use drain rods

A wire balloon should be fitted into the top of each downpipe to prevent birds nesting and debris getting down the pipe

After cleaning out gutters (or fitting new gutters), test them by pouring water into them (or use a hosepipe)

Installing a new rainwater system

The principles of installing a rainwater system are the same whatever the material, but the components you use and the methods for joining them together will vary both with the material and, for plastic systems, with the brand you use. The job breaks down into seven parts: designing the system, removing the old guttering and downpipes, repainting the fascia boards (this will almost certainly be necessary), fixing the new guttering, fixing the new downpipes, connecting to the drains or soakaway and testing the system.

Designing the system

The amount of rainwater (also known as *surface* water) with which the gutters have to cope depends on the area of the roof from which the water is collected and, of course, on the amount of rain likely to fall on it.

In this country, the figure generally taken for designing rainwater systems is a maximum rainfall of 75mm (3in) a day. To calculate the size of gutter needed to cope with this amount of rain, allowance has to be made for the fact that the wind will tend to drive more rain on to the roof than would simply fall on the flat plan area. The formula used to calculate the 'effective' area for each section of roof is: $(A + \frac{1}{2}B) \times C$ where A is the horizontal span of the slope, B the height of the ridge above the eaves and C the gutter length.

Each section of the roof is calculated in turn, which can be quite complicated if the house has a bay or a hipped roof, and allowances made for corners which are near outlets. If there are bends within 2m of an outlet, add 20% to the area for sharp bends and 10% for round-cornered bends. The gutter size also depends on where the downpipe is: gutters with the downpipe in the centre can cope with twice as much water as those with a downpipe at the end.

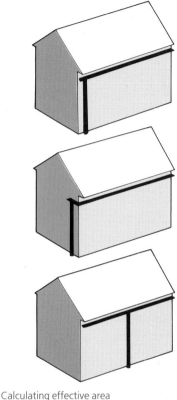

Different gutter layouts – for the same size gutter, the centre layout will cope with 20 per cent less roof area than the top; the bottom layout with 100 per cent more

Calculating effective area

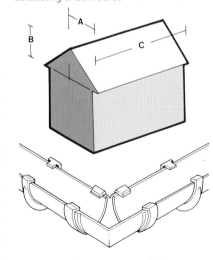

Extra support brackets are needed at external (and internal) corners.

Since semi-detached houses usually share a rainwater system, it would be sensible for both householders to do the calculations for replacing the system together. If you don't do this, it can be difficult joining different makes of new plastic guttering or joining plastic to old cast iron.

Manufacturers publish tables showing the different sizes of guttering which should be used for different sizes of house but, as a rule of thumb, 100mm (4in) or 112mm (4½in) guttering with 68 or 69mm (2¾in) downpipe, which will cope with 110 sq m of effective roof with a central downpipe (or 55 sq m with an end pipe), should be enough for all but the largest houses. The gutter will cope with more if laid with a slight (1 in 600) fall rather than being laid level.

Having chosen the size of guttering, draw up a list of the various bits and pieces you need. Working out the length of guttering can be done at ground level; calculate the length of downpipe by hanging a string from the existing gutters or downpipe. Some systems use support brackets for pipe joints, outlets and elbows; with others, the joints, outlets and elbows are themselves screwed directly to the fascia. The manufacturer's catalogue will specify the fixing distance for the gutter brackets: normally these will be every 1m (or 3ft), with extra brackets at corners and where the outlets are fitted. Guttering is usually supplied in 2m lengths, so each length needs one joint and one (or two) support brackets.

Although most downpipes are fixed vertically, there may well be other arrangements, including downpipes angled across the wall and, possibly, discharging into a hopper head. Allow for any bend you need including offsets (or a swan neck) for connecting the downpipe.

Fittings for rainwater systems

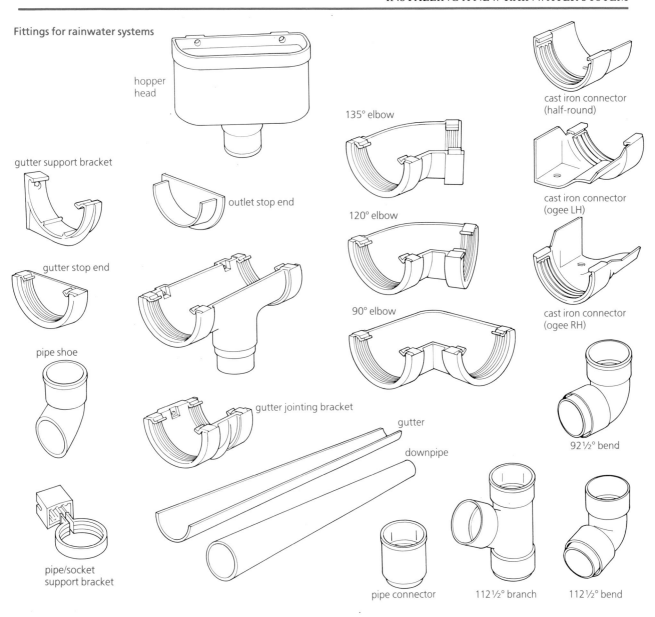

hopper head

gutter support bracket

outlet stop end

135° elbow

cast iron connector (half-round)

gutter stop end

120° elbow

cast iron connector (ogee LH)

90° elbow

cast iron connector (ogee RH)

pipe shoe

gutter jointing bracket

gutter

downpipe

92½° bend

pipe/socket support bracket

pipe connector

112½° branch

112½° bend

Removing the old guttering

Make sure that the ladder you are working on is secure and put up at the correct angle (1 metre out for every 4 metres up the wall). You might feel more secure on a hired platform tower. Choose the sort with lockable wheels since you will want to move it round fairly frequently.

When removing cast iron guttering, wear stout (gardening) gloves as the sections can be quite sharp.

Start with the length of guttering attached to the downpipe and saw off the gutter bolts (unless they can be unscrewed). Prise apart the joints with a screwdriver and remove each section in turn. You will need a rope to lower them to the ground as they are heavy – throwing them down is dangerous.

Unless the gutters are screwed directly to the fascia boards there will be brackets, either screwed to the fascia boards or screwed to the roof rafters – part of the roof covering may need lifting in order to remove them.

It's quite likely that the screws holding the brackets will have rusted in place. If so, chop through them with a small cold chisel.

Once the guttering has been removed, start on the downpipe and remove it a section at a time, starting from the top. A crowbar may help to prise off the wall brackets.

If you are removing an asbestos cement guttering system, do not attempt to cut it and don't break up the sections. Wear a face mask when removing this type of system.

Repainting the fascia boards

When you remove the old guttering, even if you find that the top part of the fascia board has not been painted at all, paint it now – after repairing any damage, including any caused by removing the guttering. Check that the fascia boards are worth repairing: if they have started to rot, you will have to replace them before starting on the guttering.

Fixing the guttering

The method of fixing the guttering varies a little from brand to brand. Make sure you get the instructions for your brand: they will usually be in the catalogue if not supplied (as they should be) with the fittings. The instructions below are for the Osma system in which the joints, elbows and outlets have their own fixing holes.

The first thing to decide is where the downpipe is going to be. This will, of course, be dictated by the position of the drain or gully and the final downpipe must run vertically downwards to this point. The position of the outlets on the fascia board can be determined by a plumbline hanging down centrally over the gully or drain inlet. Where existing downpipes are sound and disappear into, say, a concrete path, it may be easier to leave them in place and to fit the new system so that the outlet matches up with the top of the existing pipe: 68mm plastic pipe will fit into 2¼in cast iron pipe.

It is easiest to use the fascia boards as a guide to the horizontal line: check with a spirit level that they *are* horizontal or what allowance you must make if they are not.

Having established the position of the downpipe, some mathematics will be required to work out the correct height to fix the brackets. The bracket furthest away from the downpipe position will be higher than the outlet to give a fall of 1 in 500 (4mm for each 2m length or 1in for each 40ft). Choose the position of the brackets so that the end bracket is as high up as possible and the lowest point of the gutter (at the outlet position) is not more than 50mm (2in) below the roof drip.

Screw the outlet in place using 25mm rust-proof (zinc-plated) No 8 screws, fit a stop end outlet if the outlet is at the end of the roof and then move to the other end of the roof (or one of the two ends if a central downpipe is being used). Some systems have different fittings for central and end outlets; others (like the Osma) use the same fitting with an optional stop end. Taking your measurement from the fascia board, fix the bracket which will be positioned here about 150mm (6in) from the end of the run.

Now secure the plumbline to this bracket and pass the end through the outlet so that the line is taut. This is a good moment to pause and check that the outlet is correctly positioned over the drain: remember that the plumbline will now be offset by half the diameter of the downpipe.

From the outlet, fit the first bracket 1m along, against the string line (allowing for the thickness of the gutter) and fit the first length of guttering to the outlet and the supporting bracket. The gutter must be correctly positioned in the joints in order to allow room (usually 6mm) for expansion. Gutters often 'click' as they expand and contract: a little silicone lubricant on the seals will help prevent this.

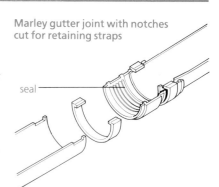

Marley gutter joint with notches cut for retaining straps

seal

In the Osma system, the gutter is pushed under the back clip of the outlet or bracket and the front clip is snapped over the gutter. The joint bracket can now be fitted to the end of the length in the same way and screwed in place. On the Marley system, the ends of gutters have a notch cut into them to fit the retaining clips at joints. Where a short length is required, you have to cut these notches yourself.

After the first length is fitted, work along the gutter in the same way until you come to within 2m of the corner of the house. Here, the length will have to be marked on the last piece when it is in position and the excess cut off with a hacksaw.

If the corner is the end of the run, mark the gutter 50mm past the roof and fit a stop end. If the gutter goes round the corner, hold the appropriate elbow (90°, 120° or 135°) against the gutter and mark where the gutter will come to inside the fitting. Remove the plumbline from the last bracket before fitting the last length, but fix it temporarily to the inside of the gutter (with something like plasticine) to provide a guide for the downpipe.

If the gutter is to be joined to existing cast iron guttering (next door's, perhaps), the appropriate connector will be needed. It is a good idea to draw the profile on to a sheet of paper when buying this. It is connected in the same way that cast iron guttering is repaired – i.e. with mastic and a new gutter bolt.

Installing new gutters

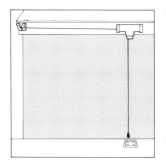

Determine position of outlet (directly above drain) and screw outlet and end bracket in place to give a fall of 1 in 500

Fix first gutter support 1m from outlet and position the first length of guttering. The hooks clip over the front of the gutter

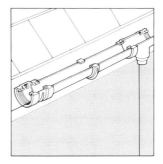

Fix gutter joint bracket, leaving the correct expansion gap, and screw it to the fascia. Continue fitting lengths in the same way

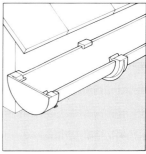

At the end of a run, measure 50mm beyond the end of the roof and cut the gutter to length. Clip on a gutter stop end

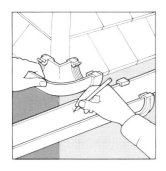

If the gutter is to be taken around a corner, hold the appropriate elbow (here 90°) in place to mark the cutting line

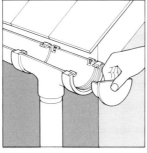

Where the outlet is at the end of the gutter run, push an outlet stop end into the free end of the outlet (or use a stop end outlet)

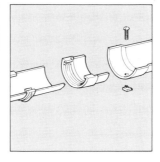

To join the new guttering to existing cast iron guttering, the appropriate adaptor must be used. Apply mastic and screw up

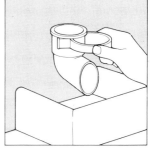

The downpipe is fitted from the bottom upwards. Push the first length of pipe into the ground or fit a pipe shoe and bracket

The backplate of the bracket holding the pipe shoe is screwed to the wall and the clip pushed into the slots

A length of pipe is pushed into the shoe and upwards to the outlet. Fix one bracket per connector and leave gaps

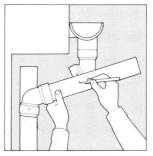

To make an offset, first push a 112½° bend on to the outlet. Hold a second bend in place to mark the length of pipe required

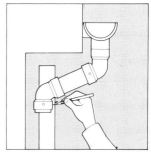

Cut this piece to length and push into the two bends to mark the correct length to cut off the downpipe. Fix pipes and brackets

Fixing the downpipes

When all the guttering is in place, you can start on the downpipes. These are generally fixed from the ground working upwards, but you must make sure that you are working to the right line if (as is common) the guttering is offset from the wall.

If the pipe disappears into a hole in the ground, simply push the new pipe into the hole: seal any slight gap around it with mastic. If the existing pipe discharges over or into a gully, a shoe supporting bracket should be screwed to the wall with zinc-plated screws (this time 38mm [1½in] No 10s) and the shoe clipped into place. See also *Connecting to the drains*. For connecting to the wall, holes must be drilled with a masonry drill bit and wallplugs fitted. Where the bricks are tough, it may be easier to drill into the mortar joints.

Work upwards in 2m lengths, fitting a pipe connector between lengths (not necessary with all systems) and a supporting bracket for each pipe connector until you reach the outlet. In the Osma system, a small hole in each pipe connector tells you when you have pushed the pipe in far enough to allow for expansion: 10mm is normally allowed.

If the outlet is directly over the pipe, simply cut the pipe to length and push it over the outlet connection. If an offset is required, use two 112½° bends and a short length of pipe. Push the bend on to the outlet and, holding the other bend against the downpipe, mark the length of the pipe required using the 'sight holes' in the bends and cut it off. Fit the second bend and mark the length to be cut off the downpipe, cut it, and assemble, fitting the bend to the outlet last. Fit a supporting bracket to the lower bend.

With some systems, you prefabricate the offset bend, solvent-welding the short length of pipe and elbows together. This requires careful

Use pencil lines to align pipes to be solvent-welded

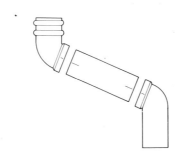

Adaptor for connecting downpipe to underground drainage

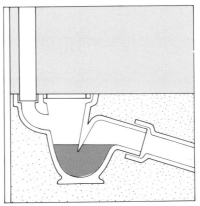

Downpipe connected to back inlet gully leading to a combined drain

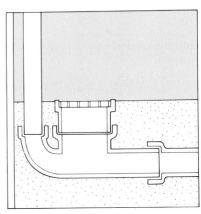

Downpipe connected to rainwater shoe (needs separate drainage)

Different ways of connecting the downpipe

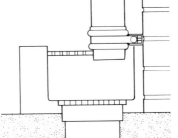

Pipe discharging below gully grid

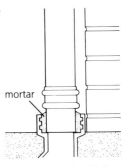

Mortared into a back inlet gully

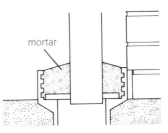

Mortared into untrapped bend using an adaptor

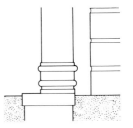

Connected to a plastic solvent-welded reducer

measurement and joining to keep the two elbow bends aligned – assemble them dry and mark with a pencil.

Where pipes are to be angled across walls, they should be cut into 1m lengths and a connector/support fitted between lengths. Use 92½° or 112½° bends for changing direction and a 112½° branch connector for joining any pipes from different outlets to a single downpipe.

Connecting to the drains

Normally, when replacing a rainwater system, the connection to the drains (or soakaway) can be made in the same way as before.

There are three main methods of connection. The first (and simplest) is where the downpipe terminates in a shoe which discharges over a gully. It would be better to make the discharge *below* the level of the grating of the gully even if it was over the gully before, to prevent leaves and garden debris blocking the gully.

The second is where the pipe is fitted to a back inlet gully; and the third is where the pipe is connected directly to an untrapped bend. With either of these two methods, the pipe can either be mortared in place (using a 2:1 sand:cement mix) or fitted with a special adaptor to seal the joint.

Testing

The simplest way to test the system is to pour water into the gutter from a watering can or run water from a hosepipe connected to the kitchen tap or a garden tap. Check all the joints between gutters and pay particular attention to bends and joints in the downpipe.

Flat roof drainage

Flat roofs are laid with a slight slope so that rainwater can be collected by conventional gutters and led down to the drains. A slight 'upstand' at one end of the roof ensures that the water will not fall off the wrong end. For garages and extensions, smaller gutters (75mm) and downpipes (50mm) can be used, and where a garden shed has guttering it may be possible to fit a water butt to collect rainwater for watering the garden. The big problem with water butts is that they can overfill if not emptied regularly and the water inside can become stagnant and a source of bacteria. These days, water butts are made of glass reinforced plastic.

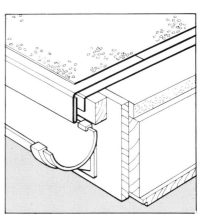

At the edge of a flat roof, the roofing felt projects over the gutter

Soakaways

Most domestic soakaways are simply holes in the ground lined with hardcore (10mm to 150mm) to within about 250mm of the top. The hardcore is then 'blinded' with sand or covered with heavy-duty

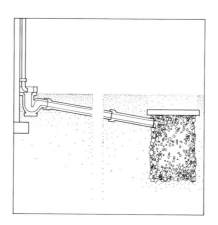

polythene and backfilled with soil to the surface. The pipe is led into the soakaway from a trapped gully fitted at the bottom of the downpipe.

The size of hole necessary for a soakaway of this type depends on the water table and the permeability of the soil and is something the local Building Control Officer will advise on: usually a hole about 2m deep and 1.2m square is sufficient.

A better type of soakaway can be constructed by making a chamber with open-jointed walling, a concrete base and an inspection cover. This has the advantages that the soakaway can be cleaned out when necessary and that you can get at it should anything go wrong. Again, the Building Control Officer will be able to advise you.

2
PLUMBING ROOM BY ROOM

CHAPTER SEVEN

KITCHEN PLUMBING

The majority of plumbing jobs in the kitchen are centred round the kitchen sink – particularly if you are putting in a new one.

Installing a new sink may well be part of a major facelift for the kitchen. The positioning of the sink in relation to other parts of the kitchen is important: it should form part of the 'work triangle' (sink, cooker and food storage) and should have a work space either side of it.

The position of the sink may well also be determined by the position of the existing plumbing arrangements. In most kitchens, it is not usually difficult to move the cold and hot supply pipes, but the position of the waste pipe can be crucial. This will probably mean putting the sink on or close to an outside wall – though it does not always have to be immediately under a window, provided that it is well lit.

Other plumbing jobs you may want to do in a kitchen include:
□ plumbing in a washing machine, dishwasher or tumble drier
□ fitting a waste disposal unit
□ installing extra water heating.
These jobs should be within the competence of most amateurs: electrical work, as always, requires knowledge and confidence, although the rules to follow in a kitchen are not as stringent as those applying to a bathroom (see Chapter 8).

Kitchen sinks

The majority of kitchens have either a white enamel or a stainless steel sink fitted on top of a base unit. These are easy to install as they are made in sizes to **'lay-on'** standard sizes of base unit.

The current trend is to have **inset** sinks which fit into a continuous length of worktop, thus doing away with the problems of cleaning either side of the sink base unit and of having to seal the back of the sink itself to the wall – it is general now for tiles on the wall to come right down to the work surface and to be sealed with (flexible) silicone sealant.

You will still probably want to have a base unit with cupboards under the sink, not least to hide the plumbing.

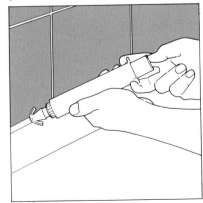

Sealing along back of sink

'Lay-on' sinks

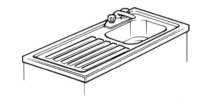

Single bowl single drainer

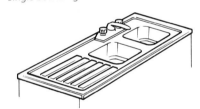

Two-bowl single drainer

Inset sinks

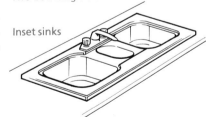

Two-and-a-half bowl

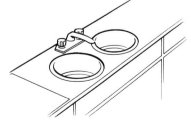

Individual bowls

Materials There is a range of sink materials. *Stainless steel* is a popular choice for both lay-on sinks and inset sinks. It has the advantage of being cheap and is both hardwearing and heat-resistant. It does, however, have a tendency to show scratches; a satin or textured finish will conceal this better than a high-gloss mirror finish. Many modern stainless steel sinks are fitted with damping pads underneath to minimise the clatter from crockery. *Vitreous enamel* sinks have a finish which is baked on to a steel base. Enamelled sinks these days are more resistant to chipping than used to be the case, and are made in a variety of colours. *Plastic* sinks have the advantage of being warm to the touch, but they may not be resistant to hot pans: *Which?* found those made from silica bonded with resin were tough and durable. As well as these main materials, you can also get kitchen sinks in Corian (a marble lookalike) and ceramic sinks in a wide range of colours, but both these types are expensive.

Sizes and shapes To be useful, a sink should be at least 180mm (7in) deep. The length and width will depend on the number of bowls and drainers it has. A look around kitchen showrooms or a glance at sink manufacturers' catalogues will show the wide range of possibilities including single and double drainers, single bowls, double bowls, two and a half bowls, and so on. The bowls themselves can be round, oval or rectangular with rounded corners. The important thing is that whichever type you choose, there is sufficient room *under* the sink to take the depth and width of the bowl and that it is possible to run the necessary pipes and wastes.

Choose the type of sink to fit the sort of tap or taps you want. You can get sinks with single, double or triple holes to take mixer taps or single taps and a 'pure' water tap.

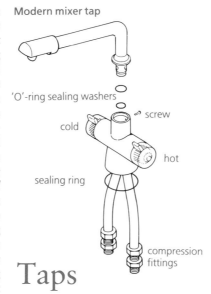

Modern mixer tap

'O'-ring sealing washers

screw

cold

hot

sealing ring

compression fittings

Taps

The main choice for kitchen sinks is between having separate taps and a mixer tap. A **mixer tap** is generally more convenient and will have a swivelling spout which can direct water to different parts of the bowl or, with a two-bowl sink, to one bowl or the other. Mixer taps for use in a kitchen (with the cold off the rising main) must have *divided flow* so that the hot and cold water do not mix until they have left the tap.

The method of fixing will vary depending on the make and design of tap. Single taps are fairly standard and are fitted in the way described in Chapter 2 – ie with a backnut and top-hat washer. Mixer taps will generally have some kind of gasket between the tap and the sink, and a rubber 'O'-ring seal between the spout and the base. Make sure that this type of tap is supplied with proper fixing instructions and that it matches the hole or holes in the sink.

You may also want to install a **rinser brush**. Sometimes, this is connected to the top of a mixer; sometimes, it has its own hole in the sink. Where the sink has two holes, a mixer tap could be fitted to one (the hole may need enlarging) and a rinser brush to the other.

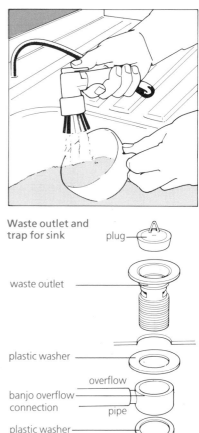

Rinser brush

Waste outlet and trap for sink

plug

waste outlet

plastic washer

overflow

banjo overflow connection

pipe

plastic washer

back nut

waste pipe

trap

Wastes

The standard type of waste for a kitchen sink is the chromium-plated fitting, 38mm in diameter. This is fitted into the hole in the bottom of the sink on a bed of mastic or sealing gasket and will take 40mm (1½in) size of trap. The waste is secured to the sink with a large nut: the threaded portion has slots in it to line up with the 'banjo' overflow connection.

Basket strainer waste

removable basket

fixing screws

sealing washer

sealing washer

washer

back nut

Some sink bowls have larger sizes of waste hole (89mm) which can be used to fit a waste disposal unit (pages 93 and 96). These can also be fitted with a *basket strainer waste* which is useful for jobs like preparing vegetables as it allows the water through but not the peelings. If it is set to another position, the flow of water is prevented as with a normal waste plug. Basket strainer wastes are fitted in a similar way to normal wastes and bought separately.

When you buy a sink, you can usually get a 'kit' with all the wastes, traps and waste pipes you need. Where one trap serves more than one waste outlet, make sure it is fitted to the outlet nearest to the drains.

Fitting a sink

The job of fitting a new kitchen sink involves removing the old sink, fitting the new sink to the worktop or on top of its base unit, and connecting up the plumbing.

Removing the old sink

The cold tap in the kitchen is likely to be connected to the rising main as the house supply of drinking water, so turn off at the main stopcock (and drain the rising main) before disconnecting the pipe from the kitchen tap, unless you are lucky enough to find a stopcock between the rising main and the tap. Other cold water supplies within the house are unlikely to be affected since there should be enough water left in the cold water cistern to cope for some time. But it will mean a lack of drinking water – so fill up kettles and jugs to keep you going or, even better, fit a temporary tap on to the pipe after it has been disconnected so that there is then no panic to get the job finished.

To disconnect the hot tap, close the gatevalve near the hot water cylinder and open the tap to drain out the water in the pipe. So that you can use hot water elsewhere in the house, fit a temporary stop end to this pipe after disconnecting.

Where the old pipes run underneath the sink, disconnect them from the taps. The tap connectors can usually be unscrewed from the end of the tap 'tails' – provided you can get to them. If not, you may have to unscrew a compression joint or unsolder a capillary joint. When removing a compression fitting, it may be necessary to saw (gently) through the olive to get the nut off the pipe; when unsoldering a capillary fitting (with a blowlamp), wrap a wet towel round the pipe further along to make sure that other capillary fittings aren't accidentally unsoldered. If the joints are too awkward to reach or if the existing plumbing is lead, simply saw through the pipes. For more about replacing lead plumbing, see Chapters 2 and 3.

Where the old taps are in the wall above the sink, they can be left in place (and operational) until the new sink is fitted. You will have to run new pipes for the new taps, which can be connected up at the end.

Once the pipes have been removed and the waste pipe unscrewed (or sawn through), the sink should come out easily. Very old earthenware sinks are heavy: removing one is a two-person job.

Once the old sink is out, remove what is left of the waste pipe: it is extremely unlikely that you will be able to use the old waste pipe so prepare yourself for filling up the hole in the wall through which it passed and making a new one in the correct position.

Fitting the sink

Before fitting a sink on top of its unit or insetting it into a worktop, the taps must be attached, together with the waste outlet (or outlets for sinks with more than one bowl). Fitting the wastes includes dealing with the overflow pipe which invariably incorporates the fixing for the end of the plug-hole chain. The waste itself should be bedded in mastic, unless supplied with a rubber gasket; use

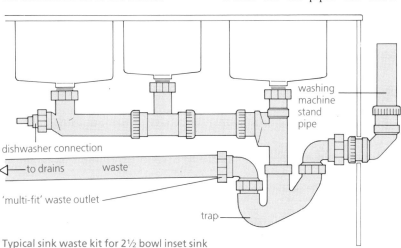

dishwasher connection

to drains   waste

'multi-fit' waste outlet

washing machine stand pipe

trap

Typical sink waste kit for 2½ bowl inset sink

'top-hat' washers for fitting the taps if the sink material is thin.

When tightening up the backnut which secures the waste outlet, hold the waste grid with a pair of thin-nose pliers to prevent it turning. Remove excess mastic.

Fitting a lay-on kitchen sink is simple: the unit is already the right size (unless the sink is 1000mm wide and the unit 42in [1067mm] wide); normally it is just a matter of screwing the fixing lugs in place.

For an inset sink, a hole of the correct size needs to be made in the worktop. Although the manufacturer will supply a template (or at least the measurements) for making the hole, it can test the nerves when cutting through an expensive length of worktop. First score the outline of the hole on to the laminated work surface, using a cutting knife fitted with a laminate cutting blade; then cut round the outline with an electric jigsaw, after drilling a hole to get the blade started off. The corners of the hole will probably need to be rounded – the job is made easier if these are drilled out first using a large bit in a wheel brace or a spade bit in an electric drill: check the sink manufacturer's instructions for the correct size. This will mean that the jigsaw can be used in straight lines from one hole to the next, reducing the margin for error (though small deviations from the line shouldn't matter). One important thing to note is that the amount of worktop left in front of and behind an inset sink can be quite small and care will then be needed in handling the worktop to prevent it breaking. Remember to seal all exposed edges of the worktop with varnish to keep out moisture.

When the hole has been made, the sink is fitted into it using either a gasket (if supplied) or a bed of mastic around the edge to make a seal. Some sinks come with the bed of mastic (or silicone sealant) already on the underside of the lip. Once the sink is in place, clips are fastened from the underneath to hold it down to the worktop.

If the sink is metal, it needs to be electrically earthed with an earthing wire run from the main earthing point (if there isn't one in the kitchen already). Most new sinks will have an earthing tag underneath – if not, you'll have to improvise by putting the earthing wire round the tap or waste outlet underneath the nuts. Also earth all exposed copper pipes.

Doing the plumbing

If you are replacing an existing sink with a new one in the same position, you are likely to find at least a cold water supply pipe in the right position. If this is in copper, you should be able to fit any extra lengths and elbows necessary to get it to the new tap position; where space is limited, using flexible copper or plastic pipe will help. Lengths of flexible copper pipe come fitted with tap connectors.

There may not, however, be a hot water pipe if the original sink relied on a sink water heater. If you want to have hot water connected to the sink tap you will have to take a supply from a convenient point and run new (15mm) pipe down to the kitchen. The main problem is likely to be getting it *below* the worktop – particularly if you are having a continuous worktop stretching from one wall to another with an inset sink. Drilling a hole for the pipe through the worktop is easy enough, but this is unsightly and will allow water and other debris through (though you should attempt to seal it with silicone sealant); or you could cut a chase in the wall to take the pipe down. This is the neater solution, but make sure that you do not bury any pipe joints: these should be accessible. Wrap pipes with insulating tape so that any mortar or plaster used in making good does not come into contact with them.

In some kitchens, the old taps may not be connected to the sink at all but be screwed into the wall above. If these are connected to old lead pipes buried in the wall, you will probably want to run new pipes to below the sink. One advantage of having pipe runs below the sink (apart from looking neater) is that it makes it easier to plumb in a washing machine or dishwasher next to the sink. The old pipes can be left in the wall and disconnected after the new pipes are installed.

The hot and cold pipes are connected to the taps by means of tap connectors: when fitting these, don't forget the fibre washer which can easily fall off and get lost. It's best to use compression rather than capillary fittings: they will be easier to connect up and remove in the future. Fit servicing valves to both pipes.

The main problem with fitting the waste pipe will be getting the position of the hole in the wall in the correct place. Many sinks are sold these days with complete waste kits which incorporate all the correct bits and pieces you need (helpful if the sink has more than one bowl and a washing machine is being plumbed in at the same time), but it will take some careful measurement to mark the position of the hole on the wall. Do this before securing the sink so that you have more room for making the hole and clearing up the mess afterwards. If you make a larger hole than you meant to, you can fill it with a d-i-y- plaster.

Fitting the waste pipe through the wall is only part of the story. At the sink end, it has to be connected to the waste trap; on the outside, it has to be taken to the drains. The inside job will not be difficult if you have got your measurements right in the first place; the difficulty outside will depend on whether there is a convenient gully into which to take the pipe. If there is no gully, you will need to connect into the soil stack – a job explained in Chapter 5.

Fitting a new kitchen sink (see page 90)

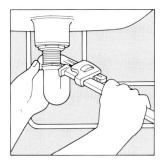

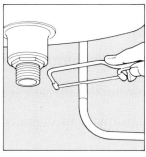

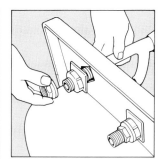

To remove the old sink, the waste pipe must first be unscrewed: position a bucket underneath to catch drips

If there are no compression fittings on the pipes leading to the taps, saw through them, trying to keep the cut square

Before the new sink is installed, the taps (or mixer tap) are fitted using 'top-hat' washers and sealing washers, if supplied

The waste will normally be bedded on a bed of non-setting mastic. Make sure the slot lines up with the overflow connection

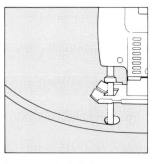

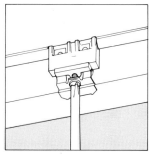

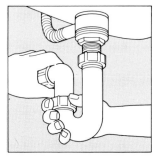

To cut out the hole for an inset sink, first mark the outline, drill a hole, score the line and cut with a jigsaw

You may need to enlarge the hole for the waste pipe or, if it is in the wrong position, make a new hole altogether

Make sure that the sink sealing gasket is in place, position the sink carefully in its hole and tighten up the clips

Connect up the trap to the waste outlet, and the waste pipe to the trap leading it out through the wall. Make good the wall

Plumbing in a washing machine (see page 94)

Connecting the water

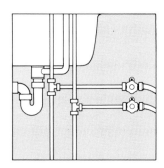

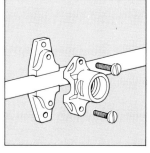

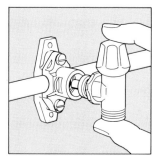

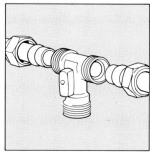

The conventional way to connect a washing machine is to tee into the hot and cold pipes and run pipes to washing machine valves

To use a self-cutting fitting, first place the saddle over the pipe at a suitable point and tighten up the screws

Now screw the valve section in by hand to cut through the pipe. Tighten with the valve vertical and attach the hose

Another method is to use a 'slip' tee fitting which contains its own stopvalve. A short length of pipe must be cut out first

Arranging the waste

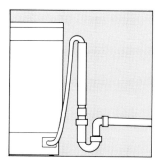

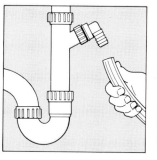

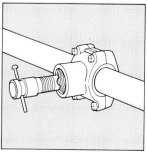

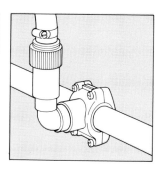

The best method for dealing with the waste is to hook the waste pipe into the top of a stand pipe fitted with a trap

If the installation instructions allow it, you may be able to fit the drain hose on to the nozzle of a washing machine trap

A self-cutting waste kit is connected to the sink waste pipe. The hole is made with a cutting tool

The coupling (incorporating a non-return valve) is connected and the washing machine drain hose secured with a clip

Installing a waste disposal unit (see page 96)

A waste disposal unit fits underneath the sink. It is provided with an electrical flex and a waste outlet

The inlet assembly and waste plug are fitted to the 89mm hole in the sink; the inlet may need to be correctly positioned

Fit the two plastic mounting rings to the bottom of the outlet – make sure there is enough room to work for the next stages

Unscrew the mounting bolts so that they do not protrude into the rim and fit the O-ring sealing washer to the seating

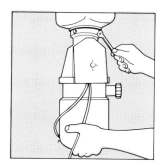

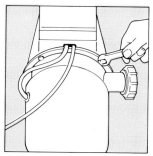

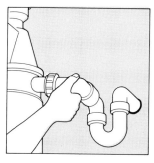

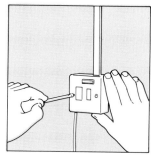

Hold the unit in place (with help if necessary) and tighten up the nuts holding the waste disposal unit in place

If necessary, loosen the body nuts and rotate the bottom of the unit so that the wastes line up and the switch is at the front

Make the hole in the wall for the waste pipe, fit the trap and the waste pipe to the unit – allow the correct slope on the pipe

The flex of the unit can be taken to a switched fused connection unit, preferably with a neon warning light – see page 96

Plumbing in a washing machine

Having an automatic washing machine more or less permanently plumbed in leaves the kitchen taps free all the time and saves having to connect up each time you use the machine. A dishwasher is another appliance you would want to treat in the same way.

Checking the machine

The majority of automatic washing machines are hot-and-cold fill – that is, they need to be connected to both the cold and hot water supplies; a few washing machines (but nearly all dishwashers) are cold-fill only. Even if your machine is hot-and-cold fill, you don't have to connect it to both supplies: it will still work if connected only to the cold (but *not* only to the hot), since the machine's internal heater heats the water to the correct temperature. It's a matter of cost and convenience: a washing machine connected only to the cold supply will take longer to complete its cycle as it has to wait for the water to be heated up (the hotter the wash temperature, the longer it will take); this will probably also cost more as you will most likely be using full-price electricity to heat the water in your machine. Most other methods of water heating are cheaper, unless you can run your washing machine at night to take advantage of the Economy 7 off-peak tariff.

The hot supply can be taken from most water heating systems, but *not* from a single-outlet water heater.

If you are connecting to both hot and cold water supplies, check the machine's instructions for the relevant water pressures which it will accept. Many machines have a removable flow restrictor for when the cold supply comes from the cold water storage cistern rather than from the rising main, as is normal in a kitchen. If you are installing a machine in a bathroom (see Chapter 8) or elsewhere on the first floor, check (by measuring) that there is sufficient 'head' from the base of the cistern to the machine. Each foot of head is roughly ½lb/sq in: each metre is roughly 0.1bar.

If you are worried that the mains water pressure for your house does not fall within the range specified for your machine (typically 0.3bar to 10bar), you can check the pressure with your local water company, but it is rarely a problem.

Before starting, make sure the machine is level and on a firm base.

Connecting the water

Provided there are water pipes below the sink, making the connections for a washing machine or dishwasher is simply a question of turning off the water, draining the pipes and fitting tee connectors into them. Special tee fittings are sold specifically for plumbing in washing machines; they incorporate a stopvalve, and have no internal stops, which makes fitting easier. There is also a special type of connector which clamps on to the pipe and makes the connection by forcing a piece of the pipe out. The action of cutting is usually by screwing in a self-cutting bit. Both connectors have a ¾in BSP fitting to take the washing machine hose.

Fitting a conventional compression tee (or a push-fit tee) has the advantage that you can then run copper or plastic pipe to the position of the washing machine (handy if it is some way away) and then fit special washing machine stopvalves (with ¾in BSP screwed outlets) in place before connecting the washing machine's flexible hoses. When running the new pipes from connections on the pipes leading to the kitchen sink, it may be necessary to bend one round the other. If the machine does not have an in-built back-siphonage device, fit a check valve.

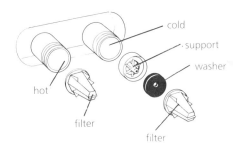

Connections to a washing machine with flow control washer for mains connection

cold
support
washer
hot
filter
filter

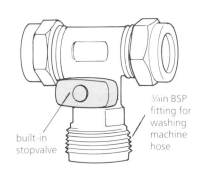

Washing machine tee valve

¾in BSP fitting for washing machine hose

built-in stopvalve

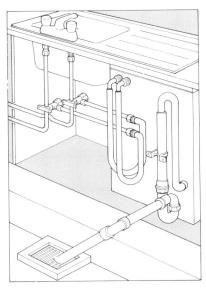

Typical plumbing arrangement for washing machine

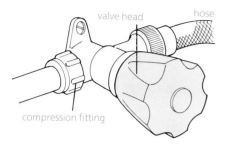

Washing machine stopvalve

valve head   hose

compression fitting

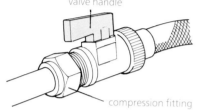

In-line washing machine stopvalve

valve handle

compression fitting

Arranging the waste

The waste from a washing machine must be connected to a trap in the same way as other wastes.

The best method is to run the hose from the washing machine to a vertical stand pipe which is then connected to a P-trap and run outside to a gully. The stand pipe must be large enough for there to be a gap around the drain hose (40mm should be sufficient) and it must be positioned so there is an 'air break' between the end of the hose and the trap.

The installation instructions will give advice on the minimum height of the top of the stand pipe: typically this will be between 500mm and 700mm (20in and 28in) off the floor.

As an alternative to making a hole in the wall, the stand pipe could be fitted to the end of a replacement trap fitted underneath the kitchen sink, provided the dimensions specified in the installation instructions can be met.

Many washing machine manufacturers specify that only a stand pipe can be used. If other methods are allowed, the drain hose (with its end

cut off) can be pushed either on to the nozzle of a washing machine trap fitted under the sink or on to the nozzle of a self-cutting waste connector fitted to the kitchen sink waste pipe. See drawings on page 93.

Electrical connections

A washing machine does not need anything more elaborate than a 13A plug to connect into a convenient socket. The difficulty is likely to be in providing a convenient socket.

In many kitchens and utility rooms, the washing machine will stand on its own (possibly with a tumble drier on top) and there will be no problem provided there is a socket within reach of the machine's flex. Where the machine is fitted below a work surface, however, the connection may be trickier since all the electric sockets are likely to be above the surface of the worktop and the work surface itself may well extend the full length of the wall.

One solution is to fit a switched fused connection unit (with a 13A fuse) above the work surface with a flex outlet box on the wall below, the two being connected by 2.5mm^2 twin-and-earth cable chased into the surface of the wall. The washing machine flex is then connected into the flex outlet box. The disadvantage of this is that the flex has to be disconnected if you want to remove the washing machine at any time.

Another solution is similar in terms of the amount of wiring and work involved, but this time a 20A double-pole switch (or fused connection unit) is fitted above the work surface and a single (unswitched) socket outlet below. A 13A plug is then fitted to the washing machine flex and simply plugged into the socket.

The fused connection unit in either example or the double-pole switch in the second are fitted to the ring circuit supplying the kitchen – a neon warning light to show that the

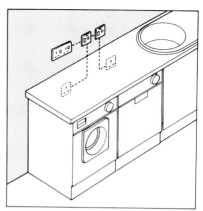

Electrical connections for washing machine/dishwasher underneath a worktop

power is connected and turned ON is a good idea.

If you are refitting a kitchen, it's not a bad idea to install a separate ring circuit just for that room. Although there is nothing in the Wiring Regulations to prevent a kitchen being supplied off the downstairs ring circuit (or even the whole house ring circuit), provided the total floor area does not exceed 100 sq m, the load put on a circuit when the kitchen is in full swing does not leave much for the rest of the circuit. If putting in a separate ring circuit, consider installing it using a *residual-current device* (also known as a residual-current circuit breaker) to provide shock protection. This is fitted in addition to the normal circuit fuse (or miniature circuit breaker) and will trip if there is a fault anywhere in the circuit, thus preventing you getting a fatal electric shock.

Testing

When all the connections have been made, turn on the water, open the stopvalves and check that the washing machine works properly and, in particular, that it will empty through whatever waste arrangement has been fitted without water spilling on the floor or flowing back into the kitchen sink.

Installing a waste disposal unit

Waste disposal installation

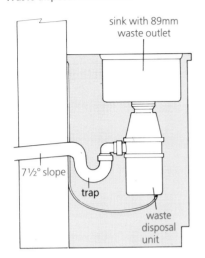

sink with 89mm waste outlet

7½° slope

trap

waste disposal unit

Most waste disposal units are designed to fit an 89mm (3½in) sink waste outlet. Many sinks will have one of these, particularly if there is more than one bowl, but don't be put off if yours has only a 38mm (1½in) outlet. Some disposal units are made to fit this size and, with a stainless steel sink, the hole can be enlarged by marking the circle round the existing waste hole and drilling a series of small holes to break out the ring of metal, finishing up with a file to make the hole smooth.

A waste disposal unit should not be fitted to a glassfibre (GRP) sink as this could crack; care should be taken fitting one to a ceramic sink.

The unit will require an electrical supply and a waste connection. The method of fitting varies slightly from one machine to another and you should always read the installation instructions before you start.

Fitting the unit

Most of the work in fitting the waste disposal unit will need to be done under the sink – clear out the cupboard before you start.

The first task is to remove the existing waste trap, by unscrewing the large nut with a bowl underneath to catch any drips, and the waste outlet, secured either with another large nut or with a single screw. Clean off any old sealing compound from the sink; on a new sink, you may need to scrape away some sound deadening material.

After checking that all the parts of the unit are present and correct, the next stage is to separate the unit into two halves, by removing the waste inlet assembly and plug from the main body of the unit.

The stainless steel sink inlet assembly is first fitted to the sink (sometimes a particular way round) and the body is then fitted to the inlet assembly. Sealing gaskets and washers should be supplied (and must be fitted): sealing compound or putty should *not* be used.

Waste disposers are quite heavy and you may need help to hold the unit while you fix it, if you can't wedge it in place. Check, if necessary, that the unit is facing the correct way once you have bolted it in place.

Connecting it up

A waste disposal unit needs a 40mm pipe connected via a trap – *not* a bottle trap. The pipe must be run with a fall of at least 7½° (130mm in every metre) to ensure adequate flushing of the system, and the waste run kept as short as possible.

The waste pipe should be taken from the trap directly out to the drains – it should not be connected into any other waste pipe – and, with a single-stack waste system, connected directly to the soil stack.

A waste disposer needs an electrical supply with a switch accessible above the worktop. See page 95

Testing

Check that no tools or small pieces of equipment have dropped down inside the unit while you were installing it. Then fill the sink about half full with water and check for leaks around the waste inlet. Remove the control plug and check for leaks in the waste pipe as the water runs away. Finally switch on.

Fitting extra water heating

It is very annoying to start the washing up in the kitchen only to discover that because someone has just had a deep bath, there is no hot water left. And, if the kitchen is a long way from the hot water cylinder, there may be a considerable wait before the water runs hot.

Fitting additional water heating in the kitchen is usually not difficult because all the services you are likely to want (electricity, gas and mains water) are all to hand.

The types of heaters you might want to fit are detailed in Chapter 4: the best type for a kitchen is an instantaneous water heater (gas or electric) or a small electric storage heater above or under the sink.

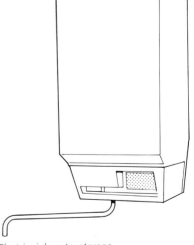

Electric sink water storage heater with touch control

CHAPTER EIGHT

BATHROOM PLUMBING

An attractive, efficient, well-planned bathroom is an asset to any home. Not only does it make life more pleasant for the family – and allow a high standard of hygiene – but it enhances the value of the house.

Bathroom planning does not call for quite the exact science of kitchen planning – there is not the same question of arranging workflows and placing activity patterns together. Nevertheless, certain groupings are a good idea. For instance, if a washbasin is put next to the bath, a shower fitting with a flexible hose can be used over both (to rinse hair over the basin); if a WC is put next to the bath it can serve as a small seat and/or table; the WC and bidet obviously go together.

Probably the most important aspect of bathroom planning is that, with a little thought and re-arrangement, it may be possible to find room for extra equipment such as a bidet, a shower or storage units in which to keep towels, toiletries and cleaning materials.

It might even be possible to relieve pressure on the kitchen by fitting in a washing machine. On the other hand, many houses have very restricted space in the bathroom and it may be necessary to think of building an extension or creating more facilities elsewhere in the house (see Chapter 9).

Planning

Many shops and manufacturers will help you with the planning, but you can have a go yourself, even if only as a preliminary to consulting the experts. Draw a floor plan of your bathroom to scale (a good scale is 1 to 20), and, to the same scale, cut out pieces of paper to represent the various items of equipment you would like to include in your bathroom. These can then be moved around until you find a satisfactory arrangement. Remember sizes of equipment vary.

Consider whether you are stuck with the present size of your bathroom. Hinged bathroom doors need quite a large floor space and it might

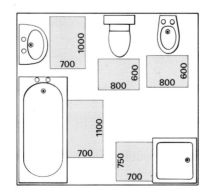

A typical bathroom layout showing activity spaces around each piece of equipment – these can overlap slightly if necessary

be possible to create a little more space by moving the door to another part of the room or by fitting a space-saving sliding door in its place. More space can be made by extending the bathroom – say by moving a non-loadbearing partition wall between it and an adjacent landing or bedroom, provided this doesn't make the other rooms too small. The ultimate answer, of course, is to have an enlarged bathroom as part of a two-storey extension with perhaps an expanded kitchen or a new or enlarged garage below.

When trying out various positions for the equipment on your plan, remember that the plumbing arrangements – particularly waste pipes – are important. Short runs with an adequate fall and as few bends as possible are desirable for waste pipes – see Chapter 5. The supply pipes do not present such a problem: they do not need much of a fall, as water is forced along them by the pressure of the mains or the head of water created by the height of the cold water storage cistern, so the supply can be taken to any part of the room.

Even so, it is good practice today to hide as many of both types of pipe as you can – inside built-in units, behind panelling and, in the case of supply pipes, under the floor, inside airing cupboards or above the ceiling, provided that joints are accessible. Some people even bury supply pipes beneath the plaster, or inside hollow walls, but this can present a hazard when later occupants, unaware of their presence, start to make fixings in the wall. In many bathrooms, the bath, basin and other equipment will be on an outside wall so that waste pipes can conveniently go to a hopper head or soil stack.

A good alternative might be to build a peninsular partition unit between, say, basin and bath in which both supply and waste pipes could be hidden. Such an arrangement would also make for economy in pipe runs. Remember that all pipe joints must be accessible for maintenance.

Note that it is often difficult to move the WC, as it cannot be positioned far from the existing soil pipe without expensive modifications to the drainage. One possibility, if you really are not happy with the position of the WC, is to install a shredder unit which breaks down the waste so that it can flow along an ordinary 40mm waste pipe (see page 116 for details).

Beware of making your bathroom *too* individual, otherwise you risk lowering the value of your house. In a cramped room, for instance, you might decide to dispense with the bath and install a shower cubicle instead – to make room for a bidet, perhaps. But such an arrangement would not be popular with everyone.

One important decision is whether the WC should be in a separate room. Many people think it is more hygienic to have the WC close to the washbasin, bath/shower and, especially, the bidet, if there is one.

Removing the partition wall between a small bathroom and WC can create extra space, but may lose some flexibility.

In a busy house, it would be best to have at least one separate WC, equipped with its own wash basin.

To separate an existing WC from

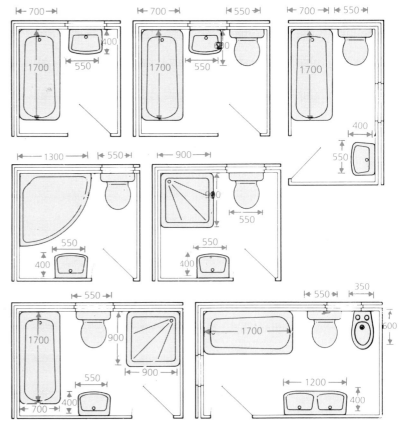

To plan your bathroom, draw a scale plan and use pieces of paper to represent the different pieces of equipment. Don't forget the 'activity' spaces.

the rest of the bathroom a simple partition wall can be built consisting of plasterboard nailed to each side of a timber framework. Such a partition should be about 75mm (3in) thick, plus skirting board. If you want extra sound insulation, you could fit two layers of plasterboard to each side which would increase the thickness of the wall by another 25mm or so. Space will have to be found for a doorway, plus room for the door to open. Bear in mind that a WC door must not open off a room where food is prepared. You should install a washbasin in the new small room, but this can be a mini-basin, even a corner model, just big enough for hand washing. If the basin is to be fixed to the partition wall, include extra pieces in the timber framework to take its fixing screws.

An even better arrangement would be to create a second bathroom – then both bathrooms could have a WC inside. The second bathroom could be *en suite* with a bedroom, to increase the house's value.

Bathrooms require adequate ventilation (a minimum of 15 litres/second of air removed) and rooms containing a WC need either a window of $\frac{1}{20}$ of the floor area or an extractor fan capable of three air changes an hour.

Areas required for:

	sq m
Bath and basin	3
Bath, basin and WC	4
Bath, basin, WC and bidet	4½
Shower, bidet and basin	4½
Corner bath, basin and WC	4½
Shower, basin, WC and bath	5½
Bath, twin basins and seating	6
Shower, basin, WC, bath and bidet	7

Washbasins

The most common material for washbasins is vitreous china, which is easily cleaned but can crack and chip. Other possible materials include enamelled steel and plastics.

Basins come with one, two or three holes, depending on the type of tap(s) you choose.

Wall-hung basins are held in place by a bracket or brackets fixed to the wall. Such a basin can be positioned at any height and it leaves the floor space below clear, but the supply and waste pipes are left fully on view. Wall-hung basins are generally 470mm to 700mm wide; 400mm to 500mm deep.

A washbasin is heavy, and can put quite a load on its fixings. The wall to which it is attached must therefore be a strong one, which usually means that it must be solid. If hanging a basin on a hollow wall (plasterboard or lath-and-plasterboard on a timber frame), you should cut away the plaster (and recess the upright timber studs) to fit a horizontal timber

Types of washbasin

wall-hung

semi-recessed

Two ways of hiding the plumbing for a basin:

in a peninsular partition

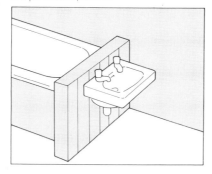

behind panelling

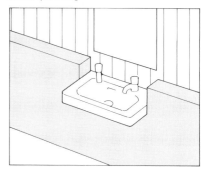

batten to support the basin. This will be concealed once the basin is fitted.

Pedestal basins cover up the plumbing, at least from the front, although it may be visible at the sides unless you add an extra fillet between pedestal and wall. The big drawback with this type, however, is that its height is fixed – generally the rim is 800mm off the floor. This is too low for many people – and too high for small children – though the pedestal can be mounted on a low plinth (with a small 'step-up' for children if necessary). Moreover, the pedestal

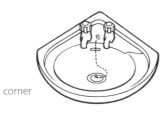

corner

pedestal

takes up floor space, and makes cleaning more difficult. The pedestal offers some support for the basin but a wall fixing is still necessary. Pedestal basins are generally 535mm to 870mm wide, 400mm to 600mm deep.

Other types of basin include the **corner** basin and the **semi-recessed** basin, which is partly let into the wall. These two are essentially space savers, and so are usually confined to very small rooms (see page 115 for details on **vanity** basins).

Fitting a new basin

A new basin must be fitted with taps, waste outlet, plug and chain (or pop-up waste) – it does not arrive with these complete. It is as well to fit these right at the start, preferably before you remove the old basin and certainly before you mount the new basin on the wall, to cut down the time your family is without its use.

All basins need an overflow so that, if the plug is inadvertently left in and the taps turned on or just dripping, the unwanted water will be carried away instead of pouring over the sides of the basin when it gets full. The overflow of a ceramic basin is traditionally an integral part of the fitting, and the slotted waste outlet is positioned so that it lines up with the opening to the overflow. The rim of the waste outlet is set in a bed of mastic (or the gasket supplied) to make it watertight, and the whole item held in place with a washer and backnut fitted under the basin to the threaded tail of the outlet. This nut must be tightened properly so that the outlet stays in place, but beware of over-tightening or you risk cracking the basin. The waste plug chain is held by a bolt and nut, fitted in a pre-drilled hole in the basin, or as part of the fixing for the overflow outlet fitting.

Like all other fittings, the basin needs a trap designed to hold a small reservoir of water that stops drain smells coming into the room. Two types are normally suitable for a washbasin – the bottle trap (shown in the drawing below) and the P-trap (right). If the waste pipe runs vertically out of the room, an S-trap should be used. The trap is screwed on to the tail of the waste outlet, with an internal washer to make the joint watertight. Sufficient water will collect in the trap as soon as the basin is in use.

Instructions for fitting taps are given in Chapter 2.

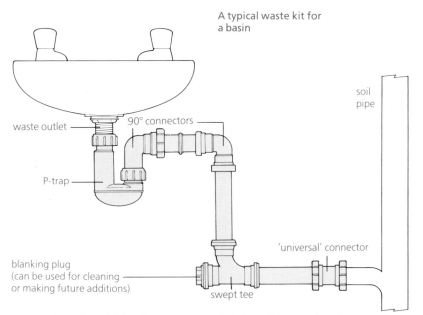

A typical waste kit for a basin

waste outlet · 90° connectors · soil pipe · P-trap · blanking plug (can be used for cleaning or making future additions) · swept tee · 'universal' connector

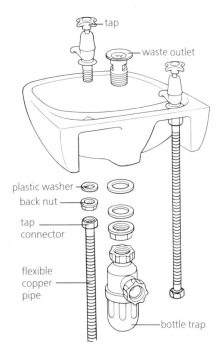

tap
waste outlet
plastic washer
back nut
tap connector
flexible copper pipe
bottle trap

Connections to a wash basin

Removing the old basin

The way in which you remove an old basin will depend to some extent on whether you wish to recycle any of the parts from it and whether the new basin is going to be in the same position or elsewhere.

Removing the pipes from the tap 'tails' and unscrewing the waste pipe from the trap is usually easy enough – provided there is room to work. If, however, you are replacing the pipes and the waste or moving the basin to a new position, it may be simpler to cut through the pipes (*after* turning off the water and draining the pipes). If you are planning to fit servicing valves to these pipes (to make future maintenance easier), you can fit them now and turn the water back on again.

What is likely to be more difficult is removing the old taps from the basin. You would only want to do this if you wanted to re-use them – in which case you probably wouldn't mind if the basin were damaged.

You may need to use some force (or a hacksaw blade) to remove the screws fixing the old basin brackets to the wall if they cannot easily be unscrewed.

Fitting the new basin

The precise method of mounting the new basin will vary according to the type and make – see page 108 for an illustrated step-by-step guide. It is important to use long rust-proof screws. Make sure the basin is fully secure, then connect up the supply pipes to it. If you have to run new pipes to the basin position, follow the guidelines for cold and hot water systems given in Chapters 3 and 4 and for working with copper (or plastic) pipes in Chapter 2. Often, of course, you will be able simply to put a new basin in roughly the same position as the old and re-use the old pipes, perhaps using a tap adaptor or an extra length of flexible copper or plastic pipe.

If you are replacing an old lead waste with the modern plastic sort, you will need to position the basin to work out where the hole in the wall is to be drilled (unless the old hole can be used): the actual connections can usually be done once the basin is fixed in place.

Make sure there is no strain on any of the pipes, then restore the water supply, and turn on the taps to test for leaks and to fill the trap.

Baths

Baths are made in three materials: cast iron (the traditional one), enamelled pressed steel, and plastic, usually acrylic. If you are putting in your own bath, plastic ones have the advantage that they are the lightest to carry and the easiest to manoeuvre during the installation.

The traditional rectangular shape is still the most common and the cheapest. The most common size is 1700mm long by about 700mm wide and about 500mm high. Shorter versions are available, however – worth considering if you have a small bathroom or want to fit an exact space. Different widths are available, too.

The very latest types of baths are *spa* baths, with rows of air holes along the bottom, and *whirlpool* baths, with side jets sending in a mixture of air and water. Both systems can also be fitted as 'kits' to existing baths.

It is a good idea to install as long a bath as you can for comfort. On the other hand, bigger baths cost more to fill. Any space between the end of the bath and the wall can be put to good use – for example, a cupboard for cleaning materials or a built-in linen basket.

The normal place for taps is at the end of the bath above the plug hole, but you can choose a bath with the tap holes in one corner or at the side.

You can get lots of extras for a rectangular bath – handle grips, slip-resistant bases and reclining backs are some of the most common. Even the cheapest baths feature these.

More unusually shaped baths mostly come in acrylic because it can be moulded so easily. Most common is the triangular bath, designed to fit into a corner. The actual bathing space of a corner bath may well be oval, with the rest of the triangle used to form a shelf or a bathing seat. The wall length of corner baths is typically 1300mm to 1500mm.

In addition, there are corner baths in which most of the shape forms the actual tub; round baths, with heart-shaped – or clover-leaf – bathing areas; and rectangular baths with oval, double or offset rectangular bathing areas.

The most obvious drawback with these, apart from their extra cost, is that they take up a lot of space. Some of them are so big you cannot get them through the average bathroom door – never mind find room for them on the floor. And if a bath is large, it takes a lot of water to fill it, which can add considerably to your fuel bills. The extra water could put a strain on an old floor although a modern floor in good condition should be able to stand the load.

Another disadvantage with such baths is that you may not be able to choose the position for the waste and tap holes and they could be in an awkward spot from the plumbing point of view.

Replacing a bath

If your new bath is a replacement for an old one, as opposed to being fitted in a new bathroom, you will discover very quickly how heavy cast iron baths are. In fact, many plumbers do not bother about removing old cast iron baths intact but smash them up on the spot with a sledgehammer. If you do this, cover the surface of the bath with thick material – sacking, old blankets or quilts – and protect any other equipment or tiles in the bathroom that you intend to keep. And, of course, protect yourself by wearing goggles, thick gloves and long-sleeved shirt and sweater. Ear plugs are essential in a confined area. Take care, too, when carrying the jagged bits and pieces of the bath outside.

There is no need to disconnect the bath from supply and waste pipes until the smashing-up is complete, by which time the connections will be more accessible. But it is obviously essential to shut off the supply and open the taps to drain off the pipes before you begin.

However, smashing up a bath can be a daunting process and you may prefer to remove the bath intact. Tackle this in much the same way as removing an old washbasin – that is, saw through the pipes rather than trying to free the retaining nuts, which will be even harder to reach in the case of a bath, and dispense with the metal waste system. Many baths

Types of bath

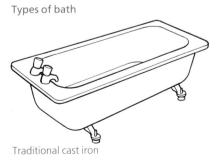

Traditional cast iron

Plastic with corner taps, hand rail and dipped front

Plastic corner bath

Connections to a bath

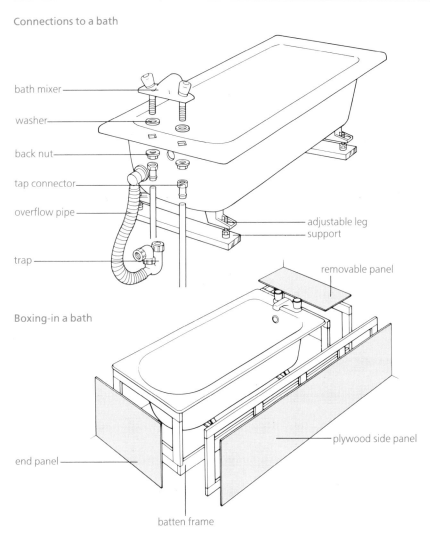

bath mixer

washer

back nut

tap connector

overflow pipe

trap

adjustable leg
support

Boxing-in a bath

removable panel

plywood side panel

end panel

batten frame

Plastic baths are supported on a cradle to prevent them creaking and sagging in use. Make sure you get the right cradle for the bath and follow the instructions for its assembly, (our illustrated step-by-step guide is on page 108). The cradle will have adjustable feet and you must ensure, by adjusting these levelling feet, that the *rim* of the bath is horizontal – its design incorporates a fall to take the water towards the waste outlet.

If your bathroom has a boarded floor, place these feet on a 25mm (1in) thick plank to spread the load – a full bath is very heavy.

Once you are satisfied with the position of the bath, connect the supply pipes. Close to the bath, joints should be compression rather than capillary since it is easy to damage a plastic bath with a blowlamp flame. You will be working in a very confined space, so tackle the one furthest away from you first – you might have to remove the overflow pipe temporarily while you do this, as it could be in the way. Bath taps are connected to 22mm pipe and, because space is limited, it is often easier to use short lengths of flexible copper (or plastic) pipe for the final fittings. The new waste pipe is 40mm plastic which is connected to the trap. Metal baths should be *earthed*: there will normally be a terminal for doing this.

Boxing-in a bath

Baths are boxed in to provide a neater installation and also to hide the plumbing and, with a plastic bath, the cradle. As a bonus, you are also spared the task of laying and cleaning floor-covering in the awkward area under the bath.

The manufacturers supply front and end panels for most baths, but often these come as an extra that can be quite expensive. You might find it cheaper to make a panel yourself – from hardboard or 3mm thick plywood. Such a panel can be decorated

have adjustable feet so that they can be put in dead level. Adjust these feet if you can, so that the bath sinks down on them, and it should free itself from the wall. Now you are faced with the problem of taking it outside. You will need at least one, and preferably two, strong helpers. Manoeuvring it downstairs can be particularly tricky.

If the old bath you are removing happens to be in plastic or steel, it can be disconnected in much the same way; but will be a lot easier to carry to the scrap heap.

Do as much preliminary work on the new bath as you can before

ing an existing one, so that the household is deprived of its bath for as short a time as possible: remember, too, that working on a bath is fiddly once it is in position.

A bath needs much the same sort of additions as a basin. However, a bath has a combined waste outlet, trap and overflow. This is positioned and then held by retaining nuts; the joints are usually made watertight by washers rather than mastic. You will find it easy to handle a plastic bath, turning it over as necessary, during this part of the job. In fact, there is no reason why you should not do the work elsewhere if space is cramped.

to match the rest of your bathroom and designed to incorporate useful space at the end of the bath.

It is a good idea to incorporate a plinth to give toe space under the panel at the front of the bath; this can be in 100mm × 25mm softwood. The front plinth is nailed to the end plinth; both their other ends are fixed to the wall or skirting board by means of small blocks screwed to their backs. All exposed wood should be painted to stop moisture getting into it. Use timber which has been pre-treated with preservative.

Frames for the panels are in 50mm × 25mm softwood and need not be planed. Those at the ends are screwed to the wall, and the front frame is nailed to the end ones. The front panel should be detachable so that it can be removed if maintenance is ever required. You could use magnetic tape fixings or touch-and-close fasteners, designed for do-it-yourself double glazing, to hold a lightweight panel in place. Or you could use plated dome head screws, (four along the top and bottom for an average bath) fixed after the final decorative treatment for the panel is in place. You can use the same method for the end frame, or nail this in place. The join between front and end panels can be neatened by a metal, plastic or timber angle strip.

Where a bath spans the whole space from wall to wall, only a front panel will be required. The frame for this should be in square timber – 38mm is a good size – and screwed to the wall at each end.

If you want to build a ledge round the bath for people to sit on, use 50mm square unplaned battens to make a more substantial frame.

You might decide to use tongued-and-grooved matchboarding instead of hardboard for the panel. This should be nailed to the frame before the panel is in position, then the assembly can be screwed in place as a whole to ensure ease of access.

WCs

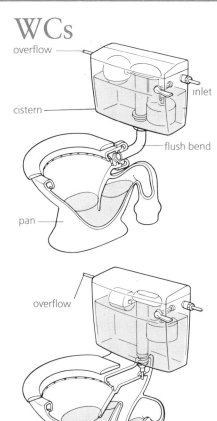

Top washdown WC with standard low-level cistern

Bottom siphonic WC with double trap and close-coupled cistern

A WC (water closet) consists of two parts – the **pan**, which is usually made of vitreous china, and the **cistern**, which holds the water for flushing, and which comes in either vitreous china or plastic. The two are normally connected by a length of pipe, known as a flush bend, but some more expensive models (called 'close-coupled') have the two parts linked.

The separate cistern is almost invariably of the **low-level** type with the cistern on the wall above the pan. But older cisterns were positioned much higher, up near the ceiling, with just enough space left above for

access, because this extra head of water was thought necessary to give a strong enough flow for flushing. Such **high-level** cisterns were usually made of cast iron, although in later years plastic was used. There are still many of them in use today and you can buy replacements, though you may prefer to replace a high-level cistern with a low-level one.

Like all other bathroom equipment, the WC pan must have a trap which holds a water seal, to stop drain smells coming into the house; this trap is an integral part of the outlet pipe. When you buy a new WC, choose the outlet to match your

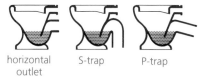

horizontal outlet   S-trap   P-trap

existing installation. If the outlet pipe disappears into the floor, the trap is said to be *S-shaped*; if it slopes only slightly from the horizontal on its route to the soil pipe, it is termed a *P-trap*.

There is, too, a choice in flushing systems. The simple kind in which the pan is cleared by the flow of water from the cistern is termed a **wash-down** WC. Some modern WCs have a **siphonic** flush which works by sucking air out of the waste pipe, thus creating a pressure differential that draws out the waste. Water from the cistern then replaces that sucked out of the pan. Siphonic flushes make much less noise and because the pan has a larger surface area of water, and water covers more of its sides, it is less likely to get soiled (but they are more costly).

Most siphonic WCs are close-coupled, and have a double trap, as opposed to the single trap standard on wash-down WCs. A double trap is more effective, although it can get blocked more easily.

Some existing cisterns may be *dual flush*, with the choice of full (9 litres) or half (4½ litre) flush: to get the full flush, the handle has to be held down. This type is no longer widely available; modern WC cisterns have a 7½ litre flush.

Both wash-down and siphonic WCs can have the cistern, the pipes and the soil pipe hidden behind a false wall, which can be of plaster-board, chipboard or plywood on a timber frame. This is not such an expensive installation as it might first appear, since a very cheap plastic cistern, without cover, can be used instead of the more costly ceramic cistern, with cover, that is often specified when the whole thing is on view. The pan itself can be floor-standing or wall-mounted.

Installing a new WC

The most difficult part of the job is connecting the new WC to the existing soil pipe. When the house was built, the pipe would have been sited so that it matched the outlet of the chosen pan. Your new WC, however, is not going to be an exact copy of the old and so its outlet and the soil pipe will almost certainly not coincide. Moving a WC is not difficult, though, since there are connectors and adaptors – notably those in the Multikwik range – to help you connect the new to the old. Take measurements of the existing installation before you shop.

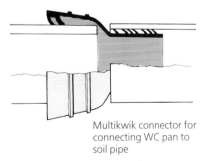

Multikwik connector for connecting WC pan to soil pipe

Before removing the old WC, give it a thorough flushing – two or three times if necessary – to make sure the pan is well washed out. Let it stand

for a while with a bleach solution in it and then flush out to clean the outlet. Turn off the supply to the cistern at the gatevalve (by the cold water cistern) before the last flush so that the cistern will not refill.

If there is no gatevalve, you will have to drain the cold water cistern: take the opportunity to fit a gate-valve – so that the water can be turned on again and maintenance in the future made easier. Now you can dismantle the old WC cistern.

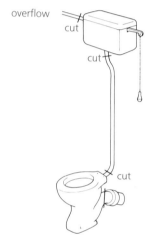

overflow

cut

cut

cut

Removing a WC – make the cuts where shown

If you are replacing an old high-level cistern that is connected with lead pipes (perhaps with a lead overflow) it will probably be easiest to saw through the pipes and the securing screws if they can't be unscrewed. On the other hand, you may want to take more care if there are any parts which you might want to re-use. Normally the easiest connection will be the nut holding the pipe on to cistern ballvalve, but if this is stiff, be careful not to exert too much force as the whole cistern could come down on your head.

Once the pipes have been disconnected, the cistern can be removed and the wall brackets unscrewed if necessary.

Now you can turn your attention

to the pan. Some water will spill when you remove it, so put down protective plastic sheeting, and be ready to mop up.

Withdraw any screws holding the pan to the floor. If there are none, you will have to push a cold chisel, bolster chisel or even a crowbar under its base and lever it up. If the soil pipe is plastic, you will now be able to pull the pan clear. Otherwise break the top bend of the trap (*not* the soil pipe) by striking it with a hammer. Then remove the pan.

The remains of the outlet pipe must be removed from the soil pipe opening: do this by chipping it free with a hammer and cold chisel. It is important that no debris falls down the drain; bung up the opening with an old rag to prevent this.

With the old WC out of the way, try the new pan in place and make sure that the connector will work. It's as well to do this when the shops are open in case you need to change it for another one.

Place a spirit level across the pan to make sure it is level. Pack it with slivers of wood or hardboard if necessary and, when you are satisfied with its position, push a bradawl through the fixing holes in its base to mark where to drill pilot holes for the fixing screws; on a solid concrete floor, mark with a pencil where to drill and fit wallplugs.

With a close-coupled WC, the cistern and the pan should normally be joined at this stage: with a separate cistern, the connection comes later.

Before securing the pan, connect the WC to the soil pipe using the connector. Once again, follow whatever instructions you are given, but in general this should be dry-jointed to the WC outlet using a rubber gasket. The WC pan can now be fixed in place with non-corroding (preferably brass) screws.

Now the cistern can be fixed to the wall. To ensure correct flushing, it

Connecting a slimline low-level cistern

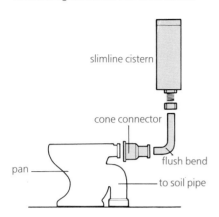

slimline cistern

cone connector

pan

flush bend

to soil pipe

Over-rim bidet

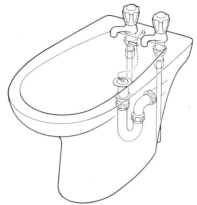

Through-rim bidet

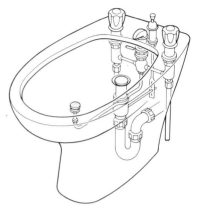

must be horizontal. The existing supply pipe will have to be adapted – i.e. lengthened or its course altered – to suit the new cistern, or you may have to go into the loft to run a new length of plastic or copper pipe down through the ceiling. Connect this pipe to the cistern's ballvalve with a tap connector. Note that WCs fed from the rising main need a high-pressure ballvalve while those fed from the cistern need a low-pressure one. Fit a servicing valve.

An overflow must also be provided for the cistern. You can use 22mm polypropylene pipe with push-fit connections for this. If necessary drill a new hole in the wall to take the overflow pipe outside.

The final task with a separate cistern and WC is to fit the flush bend. This is linked to the underside of the cistern by means of a screwed connection and to the pan with a rubber cone connector. The connector is pushed on to the flush bend, and its outer collar turned back on itself. The end of the flush bend (cut to length if necessary) is inserted into the spigot in the back of the pan and the collar is pushed back around the spigot.

Replacing a high-level cistern
If you have an old-fashioned chain-operated high-level cistern you may want to replace it with a low-level

one without replacing the WC pan itself. The problem about doing this is that the pan may be too close to the wall to fit a flush bend between the pan and the cistern. The answer here is a *slimline* cistern.

Even with these, connection may be difficult and the WC seat may not stay up when you want it to. You may be able to move the pan forward and fit a longer adaptor between pan outlet and soil pipe but an old pan may be firmly connected to the soil pipe and will be difficult to release without damaging it, which will mean a new pan as well.

Faults with WCs
The two most common faults with WCs are leaks from the soil pipe and failure to flush properly.

Leaks are most often caused by failure of the putty in an old soil pipe joint. This can be picked out and replaced with non-setting mastic.

Failure to flush is sometimes caused by too low a water level, but with a high-level WC cistern it is usually caused by dirt around the 'bell'. In a low-level wash-down cistern, it is most likely to be failure of the flapvalve washer. This is easy to replace once you have drained the cistern and removed the flushing mechanism: if you cannot get a washer of the correct size, buy one larger and cut it down to size.

Bidets

The bidet is becoming more popular in Britain, and is a useful addition to the hygienic facilities of your home. If you decide to install one for the first time, the main problem will probably be in finding space for it, remembering that it must be positioned to suit the waste pipe. There are two types of bidet – the over-rim supply kind and the through-rim supply kind. Most are floor-standing, though wall-hung models are available.

An **over-rim supply** bidet fills the bowl of the bidet just as though it were a washbasin. This type can feel a little chilly when you sit on it, but this is not important when it is being put to its other uses of footbath, basin for small children, or for soaking clothes.

A **through-rim supply** bidet is warmer in use because it sends a stream of hot water round the rim before filling the bowl. Most have a spray for douching at the base of the bowl.

The problem with through-rim bidets, however (apart from being more expensive), is that they are more difficult to plumb in, due to the water bye-laws that govern their installation. There is a risk of contamination through dirty water being sucked into the mains by back-siphonage because of the submerged spray if the bye-laws are not observed. You must inform your water company if you intend to install any kind of bidet.

An over-rim supply bidet is treated largely like a basin, but if the cold supply comes from the mains, there must be a gap of at least 25mm between the tap outlet and the top of the bidet.

For through-rim supply bidets, the rules are more complicated, but one relatively simple method is to take a separate cold feed from the main cold water cistern (this is also allowed to feed a WC cistern) and a dedicated hot water feed from the pipe leading out of the top of the hot water cylinder, which feeds no other equipment, is fitted with a check valve and has a separate vent pipe (after the check valve) leading back to the cold water cistern.

Installing a bidet

To install a bidet, begin by turning off the water, and draining as much of the system as necessary. Screw the bidet to the floor using brass screws, preferably those with a covering cap concealing the screw head. Use washers and tighten gently to make sure you do not damage the bidet. With an over-rim bidet, fit suitable tee joints into convenient hot and cold pipes supplying other bathroom fittings. The supply to the bidet will be in 15mm pipes, so if the nearest bathroom pipes are 22mm, use a reducing tee. Run the pipes to the taps of the bidet. If the cold supply comes from the rising main and you want to use a mixer, this must be of the divided flow type.

A through-rim bidet without spray can be installed in the same way, but if you are dealing with one that has a spray, you will have to run its cold supply direct from the cold cistern, connecting it up by means of a tank connector: see Chapter 3. To run a hot supply, you need to fit a suitable tee joint – probably a reducing tee

Types of shower

Push-on rubber hose

Bath/shower mixer

Shower mixer

from 22mm to 15mm – into the pipe leading out of the hot water cyclinder, as described above.

A special bidet 'set' is fitted to direct the water round the rim; this comes complete with sealing washers. Its inlets are usually compression fittings, so tap adaptors are not required. The set is, of course, a mixer, but divided-flow versions are not available, so the cold supply must come from the cold cistern: you cannot connect to the rising main.

The waste outlet for either type of bidet is made up just like that for a washbasin. Use a slotted waste to connect to the built-in overflow. A trap must be fitted – this is screwed on to the tail of the waste outlet, just like that of a basin. The waste pipe from the trap should be in 32mm plastic pipe. If you have a two-pipe waste system, the bidet waste can discharge into a hopper head or gully – it should *not* be joined to the WC waste pipe. With a single-stack waste system, it will be connected to the stack in the normal way.

Showers

A shower is refreshing, hygienic and quick as well as economical on water. The main difficulty is ensuring an adequate flow rate.

A **bath/shower mixer** replaces the bath taps; a wall-mounted **shower mixer** needs its own supplies taken from the existing plumbing. But these types can generally be used only where the cold and hot supplies both come from a storage cistern and the cold cistern must be at a height that gives sufficient pressure for an adequate spray – i.e. a head of water of 1m to 1.5m, though there are ways of improving shower flow – see opposite page.

If the cold supply comes from the mains and the hot from a multi-point water heater (or combination boiler), a special *pressure-balanced* type of shower mixer valve can be used.

One problem with connecting to the existing plumbing can happen when someone turns on a tap (or flushes a WC) elsewhere in the house, cutting down the flow to the shower. This is disconcerting if the draw-off is hot as you could get a cold shower, and potentially dangerous if the draw-off is cold as there is a risk of scalding.

One way to avoid the problem of scalding is for the shower valve to have its own cold supply direct from the cold water cistern, and not be teed-in to other supply pipes. A better but more expensive solution is a thermostatically-controlled valve which keeps the water at a pre-set temperature. Most thermostatic valves are the shower mixer type.

Installing an electric **instantaneous** shower overcomes most of these problems. You do not need a cold water cistern – it is connected to the rising main – and many makes have thermostatic temperature control in the form of a **temperature stabiliser** and a safety cut-out which will turn the shower off if the pressure drops. Pressure is not normally a problem as

The 'head' is taken as the distance from the handset to the bottom of the cistern

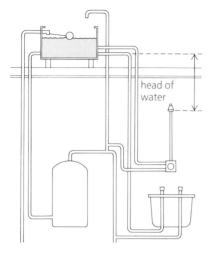

head of water

the water comes from the rising main. In fact, the limiting factor on the power of the spray is the ability of the heater to give you enough water at the temperature you have selected. The introduction of more powerful heaters – 6kW to 7kW (or, more recently, 8kW and more) – has made these showers more effective than they were when first introduced, though they will still not give as good a flow as other types of shower with a really good head of water. They are, however, always available for use.

All these types of shower can be installed over a bath or in their own separate cubicle (apart from bath/shower mixers, of course). In a bath, you can stand further away from the shower, so that, for instance, you need not get your hair wet while showering. A cubicle could be used to save space in a small bathroom, to install just a shower and not a bath. Otherwise, a cubicle is best situated elsewhere in the home – see Chapter 9.

Often you have a choice between a rigid or a flexible pipe feeding the rose. A flexible pipe is more versatile, as you can use it for hair washing at the washbasin.

Installing a shower

The method of installation varies with the type of shower.

Bath/shower mixers consist of a mixer unit with separate controls for hot and cold supplies which are mixed inside the unit. A diverter handle directs the water to either the bath spout or the shower. Most are non-thermostatic and you set the controls by bending down to turn the hot and cold taps.

Installation of these is just like fitting any other bath mixer tap, except that support has to be provided on the wall for the handset and, perhaps a rigid pipe. The handset must not be able to drop below the top of the bath.

You can fit a mixer unit to an existing bath – provided that the tap holes are in a suitable place and the correct distance apart, and that the nuts holding the existing taps are not so corroded that you cannot free them. Space under the bath will be limited and the tool to use on the nuts is a bath/basin spanner.

Shower mixers control just a shower handset so are normally used in a separate cubicle (see Chapter 9 for details) but can be fitted over the bath. A **thermostatic shower mixer** will maintain the water at the temperature you select on the dial – provided, of course, that the water in your hot cylinder is at a suitable temperature. Shower mixers are available in *exposed* versions where the valve can be seen and *concealed* where it is hidden.

The mixer is screwed to the wall or to a panel covering the supply pipes. They have hot and cold inlets which are connected to supply pipes. For non-thermostatic types, the cold supply should be run as a separate pipe from the cold water cistern; the hot supply should be the first draw-off from the pipe leading out of the hot water cylinder.

If the head is insufficient fit a shower booster pump or move the cistern

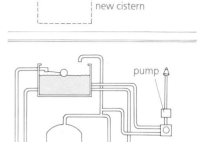

new cistern

pump

Electric instantaneous showers are particularly suitable in houses which have a direct water system or where there is an insufficient head for a normal shower. See page 110.

Improving shower flow

If you cannot get a sufficient flow of water from a shower valve, it may be possible to raise the cold water cistern to increase the head.

The alternative is to fit a **shower pump**: either an outlet pump, fitted between the shower valve and the handset or an inlet pump fitted before the mixer valve.

Outlet pumps can either be splash-proof, so they can be fitted in the showering area, or non-splash-proof, which means putting them under the bath or in an airing cupboard.

You need two (non-splash-proof) inlet pumps (or one **twin-impeller** pump), so that both hot and cold supplies can be pumped. This can give a **power shower**, with a choice of spray pattern. A power shower needs its own 22mm supplies: cold from the cold water cistern and hot from the *side* of the hot water cylinder via an 'Essex' flange.

Installing a basin

Fix the mounting bracket to the wall, making sure it is accurately positioned and firmly held in place

Fit the waste outlet lining up its slot with that of the basin (if necessary), and bedding it in mastic or the washer supplied

Install the taps, using sealing washers or mastic and fit the first length of pipe via a tap connector

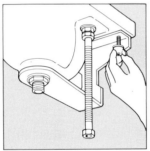

Sit the basin on its bracket and tighten the fixing nuts. Don't forget the washers and make sure it is firmly fixed

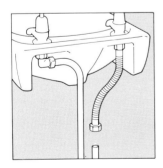

Two ways of connecting the supply pipes – with elbow and bent pipe, or a length of flexible copper pipe

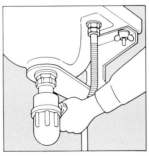

Fit a trap to the waste outlet and complete the drainage system with 32mm plastic pipe to the main stack or hopper head

Installing a bath

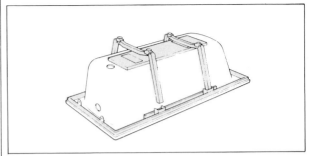

Fit the support cradle to a plastic bath by attaching the cradle legs to the board on the bottom of the bath

The cradle is also attached to the frame under the rim. A steel bath usually has brackets with adjustable feet

Before putting the bath in position, do as much as possible – for instance, you can fit the taps . . .

. . . and lengths of flexible pipe (fitted with tap connectors) together with the waste system and overflow and the trap

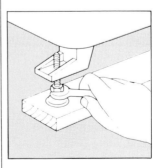

Place the bath on its timber supports, and adjust the feet so that the rim is level with the bath base sloping

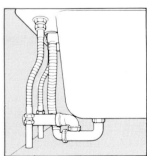

Connect the supply and waste pipes, starting with the pipe furthest away, followed by the waste pipe

Replacing a WC

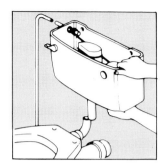

To remove the old WC, saw through supply and flush pipes; unscrew the cistern from the wall and break the pan soil pipe

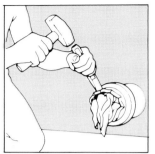

Carefully chip out the rest of the pan from the soil pipe. A cloth bung keeps debris out of the drains

Test that the new pan can be properly connected to the soil pipe, and make sure that the pan is level (pack out if not)

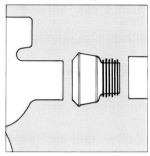

A Multikwik-type connector is used to connect the WC pan to the soil pipe. Make sure you have the correct type

Remove cloth bung from soil pipe, fit soil-pipe connector and fix pan to floor with non-corroding screws

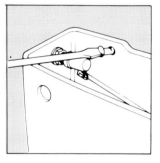

Screw the cistern to the wall, making sure it is level. With a close-coupled WC there is an in-built connector

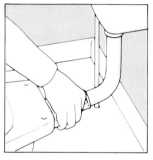

Connect the flush bend to the cistern, then to the pan. You may need to cut the flush bend down to the correct size

Connect the supply to the ball valve inlet (include a servicing valve) and fit a plastic overflow pipe, taking it outside

Installing a bidet

Floor-standing bidets are fixed with brass screws. Fit washers and tighten gently. Cover screw heads with concealed caps

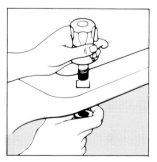

Over-rim supply taps are fitted like those of a basin. For a mixer, both hot and cold supplies must come from the cold cistern

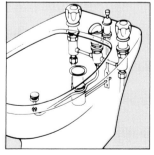

Through-rim bidets have a complex tap arrangement. They are sold complete as a *bidet set* with sealing washers

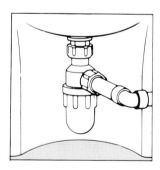

The 32mm waste can be run to a hopper head or soil stack. A trap must be fitted – there won't always be room for a bottle type

109

Installing an electric shower

Electric shower units these days should all be **splash-proof** and so can be installed within the showering area, but they should not be positioned so that they receive the direct force of the shower spray.

You will need to make decisions on how the plumbing pipes and electric cables are to be arranged close to the unit, and think about whether the pipes/cables are to be concealed (and, if so, how), whether they are to run on the surface or from the room next door.

The manufacturer's instructions will give detailed recommendations on the fitting method: one important requirement is that the shower handset should not be able to hang below the top edge of the bath or shower tray unless an anti-siphon/non-return device is fitted.

The shower unit itself will need to be partially dismantled. The backplate can generally be used to mark the position of the fixing holes: these are drilled out and fitted with wall plugs to take the fixing screws. If you need to drill through a tiled area, put masking tape on the tile to prevent the masonry drill bit sliding about. Make sure the shower unit is square on the wall.

If the electric cable from the shower unit to the double-pole switch is to be buried in the wall, now is the time to do it; also make good the wall before screwing the unit in place. Leave enough cable to connect up at the switch.

If pipes and/or cables are being run through holes in the wall behind the shower heater, now is also the time to make these holes. The first lengths of cable and pipe may need to be fitted before the unit is screwed to the wall.

While you have your electric drill around, this would be a good time to fit the wall bracket and/or sliding rail for the shower head.

The plumbing

You will need enough 15mm pipe to connect from an appropriate point on the rising main to the position of the shower. Where you make the connection depends on how exposed the rising main is and what material it is. If the rising main is lead, it would be best to get a plumber in to make the connection for you. If, as is more likely, it is copper, you can break into it at any point between the mains stopcock and the cold water cistern. The most accessible position will often be the loft: remember to insulate all exposed pipe work here.

To make the connection, fit a tee to the rising main; then run copper or plastic pipe from it, through walls and ceilings if necessary, supporting it with clips on the way. Using a compression tee may be simplest, and it is generally not difficult to cut copper pipe and fit a tee once the mains stopcock has been turned off and the rising main itself has been drained by opening the kitchen cold tap or opening the draincock.

It is essential to fit a stopvalve in the pipe leading to the shower (once this is fitted the water can be turned back on). Capillary fittings are generally neater in a bathroom, but some manufacturers specify compression fittings close to the shower and tell you not to use jointing compounds on any of the pipe fittings.

It's generally not a good idea to conceal pipework by burying it in the wall (all joints must be accessible), but you may be able to hide it behind a false wall if you are building a new enclosure or pass it through the wall from behind the shower position.

The connection to the shower unit is usually via either a compression coupler or an elbow, depending on how the pipe is led into the unit.

Before connecting to the shower unit, fit a length of hosepipe, turn on the mains (or open the stopvalve you have fitted) and flush through with clean water to remove any swarf or debris.

Once the shower has been connected, turn the water on and check for leaks before starting on the electrical work.

The wiring

An electric shower needs its own circuit which should be fused at 30A (up to 7.2kW) or 45A for wattages above this. For most showers, 6mm^2 twin-and-earth cable is sufficient, but check the manufacturer's instructions.

NOTE *Always* turn off the electrical supply before making any connections. Most of the cable for an electric shower can be positioned before making the final connection.

From the shower unit, the cable is run to a double-pole isolating switch of the correct rating (30A or 45A). Normally this will be the ceiling-mounted and cord-operated type; if the switch is wall-mounted, it must be positioned out of reach of anyone using the shower (or bath).

The switch must be fitted securely to the ceiling – either into a joist or into a supporting batten screwed between the joists. With a joist, there should be sufficient room to pass the cables down the sides into the back of the switch; with a supporting batten you will need to drill a hole to pass the cables through.

The cable should be 'chased' into the wall or run in mini-trunking. When making chases for cable in walls, run the slots vertically or horizontally, never diagonally (you may know where it is, but future occupants won't). Normally it will be possible for the cable to enter from the bottom or rear of the shower unit. Making the connection inside the

Installing an electric shower

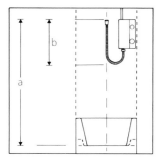

The handset should be high enough for the whole family (a); the heater should be up to 200mm below the handset (b)

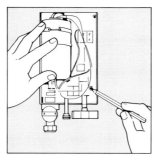

Take the cover off the shower unit and hold it against the wall to mark the position for the fixing holes and cable/pipe holes

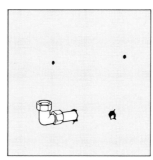

Drill and plug the fixing holes and drill out any holes necessary for passing cable/pipe through the wall

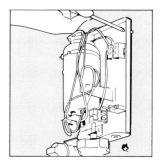

Fit a pipe elbow and a short length of pipe if necessary and secure the heater in place, making sure it is vertical

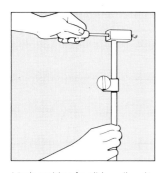

Mark position for slider rail and/or handset bracket and screw in place. Follow manufacturer's instructions for position

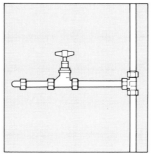

A tee connection can be made into the rising main for the water supply: fit a stopvalve to the branch supply pipe

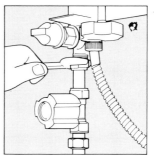

After running pipe to the shower, fit a length of hose, flush out any debris and make the final connection

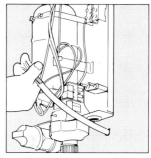

Pass a length of cable through the hole in the wall allowing enough to reach to the position of the double-pole switch

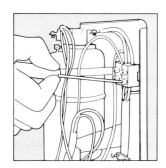

Strip the conductors and connect the three wires to the terminal block, remembering to sleeve the earth and to fix the cable clamp

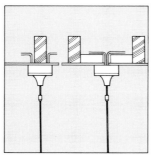

A pull-cord switch must be mounted securely on the ceiling – either to a joist or to a batten secured between the joists

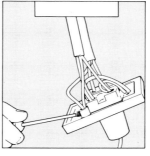

Pass the cable from the shower and a second cable (to go to the consumer unit) into the switch and connect up to the terminals

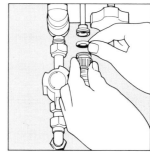

Get an electrician to make the final connection to the consumer unit. Turn on unit, run for 30 seconds, then connect handset

unit is usually obvious: the live and neutral are connected to a terminal block and the earth to the earth terminal. Note that the bare earth conductor should be covered with green/yellow sleeving both here and where it is exposed inside the double-pole switch and the consumer unit. Make sure that any cable clamp is secured and fit the cover of the shower unit.

The connections inside the switch are also straightforward: the cable from the shower unit is connected to the 'LOAD' terminals. From the 'SUPPLY' terminals, a second 6mm^2 cable is run to the consumer unit, taking the most direct route possible, clipping the cable to joists and running it in conduit where exposed.

Normally there will be a spare 'fuseway' on the consumer unit which will need to be fitted with a 30A (or 45A) fuse or miniature circuit breaker (MCB) and the cable connected to it. If not, a separate switchfuse unit must be installed – it would be best to choose one with an MCB and a residual current device (RCD). The Electricity Board will connect this unit to the mains supply, though you provide the necessary wires. Connecting to a consumer unit is fiddly and the electricity *must* be turned off. This is best left to an electrician.

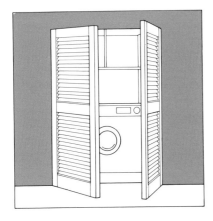

Washing machine installed behind louvre doors in a bathroom

Electrical appliances

If your kitchen is small, it may be worth considering putting the washing machine in the bathroom.

However, the IEE Wiring Regulations stipulate that a person should not be able to stand in a bathroom and at the same time touch an electrical appliance (or an electrical switch) and a bath or shower, because in the bath or shower, the damp human body is usually in direct contact with the earth via the supply pipes, and is thus at greater risk from electrical shock. That means that a washing machine and bath/shower must be getting on for two metres apart. Only in a bathroom with a large floor area could that be easily satisfied. A better solution, therefore, might be to site the machine in a cupboard, with either a hinged door or a clip-on panel, so that anyone taking a bath or shower would not be able to touch the machine.

Another problem is that the regulations prohibit the installation in a bathroom of a socket outlet, into which to plug the flex of the machine. The best way out of this difficulty is to have the machine permanently wired in, via a fused connection unit and controlled from a pull switch mounted on the bathroom ceiling. The switch would need to be 30A and double pole (i.e. controlling both live and neutral conductors in the cable, as opposed to single pole, which governs only the live). Use 2.5mm^2 twin core and earth cable, tapping into the home's existing ring circuit, probably at a convenient socket outlet. The cable should then be run from this point to the ceiling switch, then to the fused connection unit, the flex of the machine being connected to this. Preferably the cable should be unseen – under the floor, buried in the plaster of the wall, and above the ceiling. Sufficient slack should be left

in the flex to allow the machine to be pulled out for use and/or servicing.

The regulations also specify that exposed metalwork must be earthed – see page 41.

There can also be a problem with water pressure. Washing machines need a fairly strong pressure in order to operate properly. In a first-floor bathroom, pressure will not be as strong as in the kitchen on the floor below because the head of water is not so great, so a machine that will work on the ground floor might not do so effectively upstairs. You might be able to connect a machine in your bathroom to the rising main, but if it requires a hot as well as a cold feed – and most do – this will not get you out of your difficulty. Chapter 7 gives details about water pressure, but a rough and ready test is to turn on a tap in the bathroom, and put your finger over the outlet. If your finger is forced away, the pressure should be strong enough.

There is one other difficulty. The bathroom in a small house will probably have wooden floorboards, as opposed to the solid concrete floor in a ground-floor kitchen. This can have two results. First, the vibration can cause the machine to move around. Secondly, the motor can make a loud noise on such a floor, especially on the spin cycle.

The way to deal with these problems is to mount the machine on a piece of 19mm (¾in) thick chipboard screwed to the joists (not the boards) of the floor (the position of the floorboard nails will show where the joists are). The chipboard must be absolutely horizontal in both directions – check with a spirit level and pack out with small pieces of wood or washers as necessary. The chipboard will reduce noise levels, especially if rubber grommets or washers are fitted to the screws between chipboard and floor.

CHAPTER NINE

PLUMBING IN OTHER ROOMS

A single bathroom and WC are often under pressure in a family house so extra facilities elsewhere can be very useful, especially when guests are staying or if a disabled or elderly member of the household cannot easily use the main bathroom.

The best solution is to install a second bathroom. Using suitable space in other rooms can be the answer – a basin in a bedroom can take a lot of strain off the main bathroom – but if you can find room in a bedroom for a shower, and perhaps a WC, too, this will cut down early morning queues.

There may also be a solution downstairs: by converting an existing WC, by fitting a shower in a utility room or even by building on a small extension.

Design considerations

When choosing a place for extra fittings, it is important to consider the plumbing arrangements. Most houses have all their pipework at one end or corner of the house, and plumbing away from this 'water end' may mean long supply pipe runs and difficulties with wastes. Often the practicalities of arranging the waste connections will determine where you can install your new fittings.

Wastes

The main problem with waste pipes is the limit on their length. If you have a single-stack drainage system, basin waste pipes leading into the existing soil stack are not allowed to be longer than 1.7m using 32mm pipe or 3m using 40mm pipe (see Chapter 5 for more details). Where there is a two-pipe waste system, wastes to a gully or hopper head can be longer, but excessively long pipe runs should be avoided. In either case, if the new fittings are to be installed a long way from the existing drainage points, a new soil stack will probably be required. See Chapter 5.

Wastes are generally easiest and neatest to arrange where the new fitting is installed against an outside wall and the waste can go straight out through the wall. Where a waste pipe has to run inside, you will usually have to run it along the skirting and box it in later, unless you can position it under the floorboards.

The length of the soil pipe from a WC to a soil stack is not usually critical (the maximum length is 6m), but the pipes are large and cannot easily be taken under the floor so you will be limited as to where you can install a new WC. If the best position for it is not convenient for the existing soil stack, you will either have to install a new soil stack or else put in a motorised shredding-and-pumping unit (such as the Saniflo), which will reduce the contents of the WC to a slurry and discharge it to the soil stack through a 22mm waste pipe, provided an approved fitting is used – see page 116

Supply pipework

In comparison with fitting the waste pipes, laying on a supply of water to a new fitting is relatively simple. For a new basin (or bath) to be installed upstairs, the hot water supply can be a branch from the bathroom hot water pipe or from the pipe coming out of the top of the hot water cylinder – the vent pipe. It may be most convenient to make this connection in the roof space, but if you do this the pipe must be taken off at a point lower than the base of the cold water cistern. For a shower mixer valve, the hot supply should be the *first* connection from the top of the hot water cylinder.

Where hot water is supplied by a multi-point water heater (or combination boiler), you can join into the pipework anywhere that is convenient. Note that only a *pressure-balanced* shower may be used (with cold supplied from the mains).

The point from which you take the cold water supply will depend on the type of fitting. The cold supply to a basin or WC can be teed off the bathroom cold water supply pipes or taken directly from the cistern. The cold supply to a shower unit is usually best taken directly from the cistern as this avoids the problem of scalding when other fittings are used. (If you are fitting an instantaneous electric shower, it must be connected directly to the rising main.)

If a new basin or WC is to be on the ground floor, it will often be simpler to arrange the cold water supply from the rising main or a branch off it (check with your water company that you can do this). The hot water supply can come from the pipe feeding the kitchen hot tap.

There are no restrictions on the length of cold water pipe that you can have, but hot water pipes are best kept as short as possible and must be well lagged to reduce heat loss. The maximum recommended length of 15mm hot water pipe is 12m. If the distance from the hot water cylinder to the fitting is more than this, it would be better to use a separate water heater. The maximum recommended length for 22mm pipe is 7.5m

Consider using polybutylene or polyethylene pipe for long runs to new fittings. Its flexibility means that it can easily be bent round obstacles or threaded through holes in floor joists, and the long lengths in which it is sold enable you to make uninterrupted pipe runs.

If pipes under the floor have to cross joists, use a spade bit in an electric drill (or a brace and bit) to drill holes at the centre of each joist for flexible pipe or cut a notch for copper pipe (no deeper than 18mm for 15mm pipe).

Remember to support all your pipework properly, and to lag all hot water pipes and any cold water pipes that run through lofts and other unheated spaces. Fit servicing valves to make rewashering and other maintenance easier in the future.

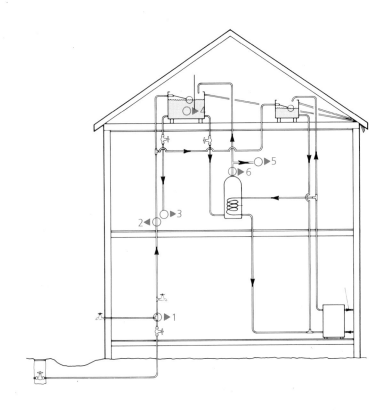

Ventilation

Wherever plumbing fittings are installed, it is a good idea to fit an **extractor fan**, perhaps the small unobtrusive type intended for ventilating WCs, to remove moisture-laden air and so reduce condensation. Showers in particular release a lot of moisture to the air, even when there are opening windows nearby. The fan switch must be the pull-cord type or be beyond the reach of anyone using the shower. Automatic control by a humidistat which switches the fan on when it senses a certain level of moisture in the air is useful.

If a bathroom or shower room is created in a space without windows that open, an extractor fan is required to meet Building Regulations. The fan should be wired to come on automatically whenever the light is switched on and run for 15 minutes after it has been switched off.

Hot and cold supplies

1. Cold supplies can be taken off the rising main for some downstairs fittings
2. An instantaneous shower is connected directly to the rising main
3. Most other cold supplies can be taken from the existing feed from the cistern
4. Some fittings (a non-thermostatic shower, for example), may need their own cold supply
5. Most hot supplies can be taken from the existing hot water supply pipe
6. Hot supply for a shower should be the first connection off the safety open vent pipe

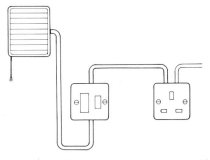

Connections for extractor fan (non-automatic)

Installing a bedroom basin

Any type of washbasin can be fitted in a bedroom, but pedestal and wall-hung basins can look a little clinical and a better choice may be a basin set into a vanity unit. Purpose-made vanity units are generally made of melamine-faced or veneered chipboard. There are both rigid (factory-made) and self-assembly versions. Alternatively, you could construct your own or adapt a piece of furniture – an old wooden washstand or kidney-shaped dressing table, for instance. If you do this, secure the furniture to the wall with angle brackets. Think about incorporating a vanity unit in a row of built-in cupboards; with a door fitted in front, the basin will be hidden when not in use.

Not all vanity basins fit into a hole: some 'lay-on' a suitably-sized unit. Semi-recessed countertop basins are now popular – often as part of a range of fitted furniture.

From the plumbing point of view, the easiest and neatest place to install a bedroom basin is more or less back to back with the bathroom basin – something worth thinking about when you are refitting a bathroom. The supply pipework can be teed off the bathroom basin pipes and taken through the wall directly to the new basin. The new waste can run straight through the wall and then in parallel with the bathroom basin waste pipe. (It is possible to join the two waste pipes, but on the whole it is better not to do so.) Where plumbing on a shared bathroom wall isn't feasible, it's usually simplest to install the new basin on the outside wall closest to the drains.

The steps that follow take you through the stages involved in plumbing a basin in a vanity unit on an outside wall of an upstairs bedroom not too distant from the existing soil pipe or hopper head.

Once the pipe runs have been planned, the detailed work involved in fitting a vanity basin in a bedroom is little different from that for a conventional basin in a bathroom – see page 99. The main variation is that vanity basins are not screwed to the wall and are often thinner so it may be necessary to fit 'top-hat' washers to act as spacers between the underside of the basin and the tap backnuts. Some vanity units are supplied with the basin already fitted; with others you usually have to cut the hole yourself. A template pattern is usually supplied to help with marking out the hole; when this isn't the case, use the basin itself as the pattern piece. Take care when marking out that the front edge of the basin is straight with the front edge of the vanity top. An electric jigsaw is the best tool for the job, but if the vanity top is laminated, cut from the back surface to reduce chipping (though any chipping should be covered by the edge of the basin). Drill a hole to start the jigsaw off.

Many vanity basins come with a and the top. Where no gasket is supplied you will need to place the basin on a generous bed of non-setting mastic sealant. It is also a good idea to paint all exposed chipboard and edges with a couple of coats of primer to seal them – chipboard absorbs water readily and will swell out of shape.

An old washstand can be adapted to take a vanity basin

Vanity basins

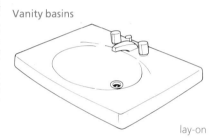
lay-on

semi-recessed

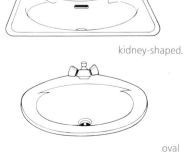
kidney-shaped.

oval

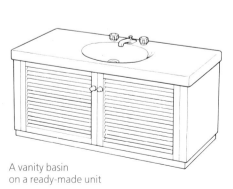
A vanity basin on a ready-made unit

Installing an extra WC

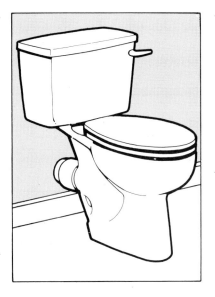

The two types of WC: close-coupled (above) and washdown (right)

If an extra WC is intended for general use by all members of the household and any visitors, it is best to find a space which is accessible from a hall or landing rather than through a bedroom.

The doors of rooms containing a WC should not open into a room where food is prepared (or where washing up is done); the room itself should contain a washbasin.

For the WC itself, you need a space around 600mm (24in) deep (from front to back) and around 550mm (22in) wide; to use the WC you need a further 600mm (24in) by 800mm (32in) in front – see page 98 for more details. If the WC is installed in a separate room or enclosure, you may need to allow for the opening and closing of the door. Alternatively, you could fit a sliding door or arrange for the door to open outwards. Rooms without opening windows (of at least $\frac{1}{20}$ of the area of the floor) must be fitted with an extractor fan wired to come on with the light switch.

WCs need only a cold water supply and this is easily arranged off a cold water pipe supplying other fittings, off the rising main or direct from the cistern. If the WC is allowed to be connected to the rising main, it will need a **high-pressure** ballvalve – see Chapter 3.

Arranging the waste is fairly easy when the new WC is installed upstairs within easy reach of an existing uPVC soil stack. Soil pipes from downstairs WCs have an S-trap and are generally connected directly to the underground drains, and installing a new one involves digging up the ground and laying underground pipes. In many situations, a much simpler solution to the WC waste problem is to fit a motorised shredding-and-pumping unit, but there must be another WC elsewhere connected conventionally.

Fitted behind a conventional WC suite, these units are quite unobtrusive and, although not totally silent, in most situations the motor runs for only 10 to 20 seconds after flushing.

Shredding-and-pumping units

Shredding-and-pumping units are not cheap, but they compare favourably with the costs of a new soil pipe and they enable you to install a new WC a long way (20m) from the existing soil stack. There are versions which can pump longer distances (50m) or which will pump up to 4m *vertically* (useful for WCs in converted cellars) and which can be used to pump the waste from other bathroom fittings as well – a basin or a bath, basin, bidet and shower.

To fit a shredding-and-pumping unit, you need a WC with a horizontal outlet. There must be a space of 190mm ($7\frac{1}{2}$in) behind the spigot of the bowl. With a close-coupled WC, this may mean that the cistern is not against the wall and may require some packing to allow the cistern to be secured or the construction of a false wall. The water supply is arranged in the usual way, but the waste is run from the unit to the soil stack in 22mm ($\frac{3}{4}$in) copper or uPVC pipe with a fall of 1 in 200. The unit has a vent connection, the top of which must be above the rim of the WC pan and can terminate either inside or outside after passing out through the wall.

The motorised unit also requires an electrical supply. This can come from the ring circuit but must be connected via a fused connection unit set out of reach of anyone using a bath or shower if it has a switch.

The Saniflo shredding-and-pumping unit needs a space of 190mm behind the WC spigot

Installing a vanity basin

Put the vanity unit temporarily in place. Mark the positions for the wall brackets. Drill holes and fit wallplugs and paint bare wood

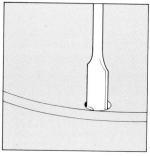

Mark the basin outline on the vanity unit top. Drill a hole inside the line and use an electric jigsaw to make the cut

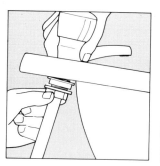

Fit the taps to the basin, using top hat washers if necessary. Fit the waste and overflow outlets. Screw on the waste trap

Set the top on the vanity unit and locate the basin temporarily. Measure up and prepare for the waste and supply pipes

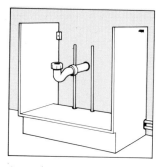

Screw the vanity unit to the wall. Install the waste pipe and get the supply pipes ready to connect at either end (P-trap shown here)

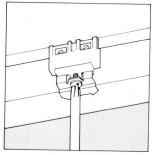

Fit the top and screw it to the vanity unit. Fit the basin on a bed of mastic or its seal. Tighten retaining brackets

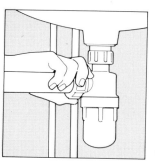

Connect the waste pipe to the trap (here a bottle trap) and pour water in to check for leaks. Connect supply pipes to taps

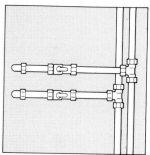

Turn off and drain the hot and cold water supply. Tee in the new pipes. Turn the water on and check for leaks

Installing a shredding-and-pumping unit

Connect the WC bowl to the unit and place the cistern on to the bowl. Try the whole set-up in position and mark the outlines

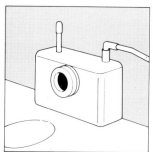

Set the unit in place and plumb the waste pipe to the soil stack at a slope of 1 in 200. Connect the vent pipe

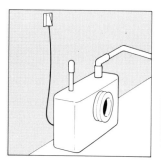

Connect the unit to the socket outlet circuit by means of a fused connection unit and flex outlet connector

Install the WC and cistern with packing if necessary. Make sure that the bowl and unit are firmly connected

Installing a shower

There are all sorts of places where a shower can be installed. In a bedroom, there is often room for one in a corner. Like a vanity basin, it can be disguised in a row of cupboards. There may be room in a landing alcove, in a space partitioned off a large airing cupboard, in the cupboard under the stairs, or in a utility room. Putting a shower downstairs is a good solution when the cold water cistern isn't high enough to give a powerful shower upstairs, though it is often more convenient to fit an electric shower connected to the rising main.

For a compact shower, you'll need an area at least 700mm (28in) square for the shower tray and a further 700mm by 750mm outside for a drying space. Where there is enough room, a shower space 1200mm (48in) front-to-back by 800mm (32in) wide allows unrestricted movement and allows ample room for a shower seat. Seats are especially useful for children or elderly people. In a large cubicle or enclosure, the tray itself need only be 800mm (32in) square with the floor beneath the seat tiled to let water run off into the tray.

Making your own shower enclosure usually means building a false wall and tiling the surround, so in many places it will be simpler to install a free-standing cubicle or to construct a cubicle against a wall. However, a tiled surround has the advantage that you can make it any size you want. If you decide to build one, take the tiles to at least 1.8m (6ft) up the wall and use waterproof adhesive and grout. If the floor beneath the shower is timber or if any of the walls of the shower are timber stud partitions, it is best to use a *flexible* waterproof adhesive.

Shower cubicles

Freestanding cubicles are usually made of plastic or glass. Many are supplied as complete units incorporating the shower itself. They tend to be expensive, but are simple to install, needing only to be connected to the supply pipes and the waste.

Making your own shower cubicle is considerably cheaper than a ready-made cubicle and is a good choice when the shower is to be installed against walls which can frame the shower area. Two side panels and a door are used where there is only a back wall; one side panel and a door where the shower is installed in a corner. The alternative in a corner is to use a two-sided sliding corner door.

Cubicle side panels and doors are generally fitted with plastic or glass. If you choose one made of glass, this should be toughened (tempered) safety glass, marked BS 6206. Glass can be clear, opaque or with a white obscuring pattern.

Choose the type of shower door which will give you the maximum opening. Sliding doors and doors which are hinged to open inwards prevent drips on the floor in front of the shower. If you choose a door that opens outwards, get one fitted with a drip channel along its lower edge. Check that the door sill or sliding track will be easy to keep clean. Magnetic closing strips help to keep the door tightly closed and make the cubicle watertight.

An alternative to a shower door is a curtain. (You could also use a curtain on a curved track all round the shower area instead of a cubicle.) The main advantage of curtains is that they are considerably cheaper than panels. It's best if curtains are weighted (with lead tape or buttons sewn into the hem); they should be designed to hang *inside* the shower tray.

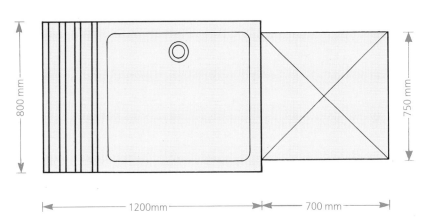

Finding space for a shower

seat

tiled area

shower tray

plinth

800 mm

750 mm

1200mm

700 mm

Shower cubicle

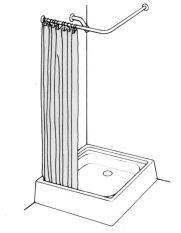

Shower tray with curtain

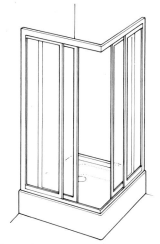

Corner shower

Shower trays

Shower trays are made in ceramic, acrylic and, less commonly, pressed steel, in colours to match basins and baths. Acrylic trays are cheaper and considerably lighter than ceramic ones. Acrylic also feels warmer to the feet when you first step into the shower; on the other hand, ceramic will last well and is more solid than acrylic, which needs support all over.

Most shower trays are square with sides between 700mm and 900mm long. Choose the largest you can fit into the available space to give plenty of room to move about. For corner settings there are trays with a curved front; remember that they will need a similarly shaped cubicle. If you plan to tile the walls around the shower tray, look for one with a special tiling lip.

Doing the work

Start the installation by running the water supply pipes to the place where the shower will be fitted; then fit the shower itself. It's easiest and least messy to do this before the tray is put in place, especially if the work involves chasing the wall to take the pipes. Note that mortar or plaster can corrode copper pipes in time and a better material for filling chases that carry pipes in walls is polyurethane foam expanding filler. Cut it off flush with the wall and tile over the top. Do not have any pipe joints buried in the wall.

Put the shower controls on a side wall of the cubicle where they will be easy to reach from outside. The plumbing involved is no different from that needed when a shower mixer is installed over a bath – see Chapter 8. The steps that follow concentrate on fitting the cubicle and the tray with its waste.

Think first about how the waste will run. If you raise the tray on a plinth you may be able to arrange the waste more easily and you will not have to work in the space below the floor. The plinth can be constructed from softwood 150mm (6in) high by 50mm (2in) wide joined with plastic joint blocks. One side needs to be made removable to allow access to the trap for cleaning, though you can get trays where the trap is accessible from the top.

If you intend to cover the plinth with tiles, allow for the thickness of the covering when you measure up. The plinth will fit more neatly against the wall if the skirting board is removed. Paint all bare wood.

Most shower trays are supplied with the waste outlet already in place, but you will need to fit a trap. This should have a 75mm seal for a single-stack drainage system or a 50mm seal for a two-pipe system, although, in practice, most showers will work satisfactorily with a shallow trap. It may be a P-trap or S-trap depending on the pipe arrangement. Fit the trap to the shower tray early on and use it as a guide for the position of the hole that takes the waste pipe through the wall. Getting the waste run right is important; it's worth checking that all is well before you finally screw the tray in place.

When you've set the tray in place and connected up the waste, test for leaks with a bucket of water before going on to tile the side walls. Take care when you are tiling to keep the grout lines exactly vertical – if they are out of true they may make the cubicle look crooked.

If you are constructing your own shower cubicle, you start with the side panel or panels: there is usually some adjustability to take account of walls that are slightly out of true, but it will sometimes be necessary to pack them out. You may find that bedding the panels on a bed of sealant will help to give a waterproof seal. In any case, it will be necessary to finish off the job by filling all the joints between the panels, the tray and wall with silicone or other sealant.

Installing a shower

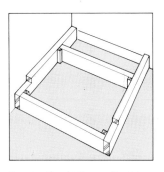

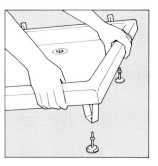

Remove the skirting and build a plinth from 150mm by 50mm softwood. Incorporate a step at the front edge if required

Run the supply pipes (including isolating valves) and connect up the shower mixer. Put it on a side wall for easy access

Fit the waste trap to the waste outlet of the shower tray and position the tray against the wall and on the plinth if fitted

Adjust the feet until the tray sits level. Mark round the feet and any wall brackets. Work out the waste pipe run

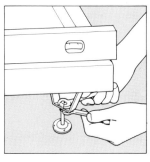

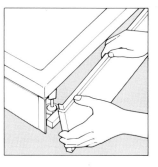

Cut the hole through the wall for the waste pipe. Cut back the plaster round the brackets. Fit the feet to the floor

Relocate the shower tray on its feet and check that it is level. Tighten the securing nuts and screw the tray to the wall

Connect the waste fittings and test them with a bucket of water. If all is well, fit the side panels of the plinth

Make good the plaster over the wall brackets and the shower pipes. Tile solid walls which surround the shower area

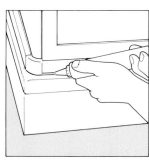

Mark, drill and plug the wall for the side panel uprights. Screw them in place. Adjust or pack out if the wall is not flat.

Adjust the side panels to fit the shower tray. Lift on to the tray. Check for level and secure in place

Prepare and fit the door panel. Test that it opens smoothly: if it doesn't, it may be necessary to readjust the side panels

Use a flexible waterproof sealant to seal the joins between the side panels, the wall and the shower tray

CHAPTER TEN

OUTSIDE PLUMBING

Most houses need an outside tap. It's useful for filling buckets for window cleaning and car washing as well as for watering the garden. In larger gardens a tap further away from the house could be useful, especially if fitted next to a greenhouse.

Fitting an outside tap is not difficult, but having one may affect your water rates and you must give your water company five days written notice of your intention to fit one.

Installing an outside tap

An outside tap is usually installed on a pipe which supplies water at mains pressure. This can be either the rising main itself or the branch off the main which feeds the drinking water tap at the kitchen sink. The pipe to the sink is often more convenient as it is likely to be on an outside wall facing the garden so that only a short branch is needed to the position of the new tap. The most difficult part of the job is making the hole through the outside wall.

Choose a position for the hole at least 350mm horizontally from the pipe you will be breaking into and about 150mm above the proposed tap position. The 350mm distance is to make room for a stopvalve and a double-check valve to be fitted in the branch to the outdoor tap; the 150mm and the stopvalve allow you to cut off the water supply to drain the tap in cold weather. The alternative is to fit a draincock below the tap in the pipe leading up to it.

It is possible to make the hole with a cold chisel and club hammer, but a masonry bit in an electric (hammer) drill makes a much neater hole. A 16mm (⅝in or No 26) bit will make a hole the correct size for 15mm pipe. If your own drill isn't up to the job, hire an industrial one with a bit long enough to pass right through the wall. If you use a shorter bit, you will need to drill from both sides of the wall and the· holes will need extremely careful lining up.

Garden taps are the 'bib' type which simply means that they have a horizontal inlet, instead of the vertical inlet of the pillar type of tap almost always used indoors. Bib taps

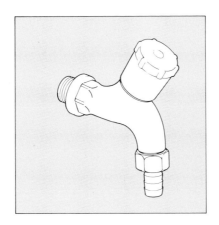

screw into a wallplate fitting which is screwed to a wall or upstand (a stout wooden post, for example). For a single outdoor tap, use a wallplate elbow fitting: if you plan to install a second tap further down the garden, use a wallplate tee fitting.

Most garden taps are brass with a simple tee handle, but there are also plastic ones with round heads. Any tap must be an 'approved' fitting (i.e. made to BS 1010): choose the type which has the handle angled away from the wall, so that you can turn it without grazing your knuckles. A threaded outlet to take a hose connector is a useful feature.

The steps on page 123 assume that you'll be using ordinary copper pipe and a conventional equal tee for connecting to the rising main. You may find it easier (but more expensive) to use flexible copper pipe for some of the lengths, though you will need rigid pipe where the stopvalve is connected. You can buy garden tap 'kits' with self-cutting valves (see page 94) with a BSP outlet to take a flexible hose. These can be connected without having to drain down the pipe or to cut through the rising main.

The job can also be done using any of the other pipe materials (see Chapter 2), but note that all pipe outside should be fitted with weatherproof insulation. The double-check valve is a vital component to prevent dirty water from the garden (perhaps containing chemicals) from being siphoned back into the mains water supply. It is fitted in the pipe between the isolating stopvalve and the garden tap.

A tap down the garden

There are two main choices for the pipework down the garden. You can either carry out a proper underground installation in copper or blue polyethylene or you can have a (more temporary) surface installation of black polyethylene pipe.

Whichever method you use, ensure it is possible to drain exposed parts of the pipe in the winter.

If you opt for copper, you will need to use the underground grade – BS2871 Table Y – and buy it in lengths long enough to eliminate joining underground: copper can be bought in lengths up to 6m or in 20m coils if 'dead-soft' temper copper tubing is used; blue and black polyethylene are available in much longer coils (50m and more). If

copper pipe has to be joined underground, capillary fittings may be used, or the manipulative type of compression fitting but this requires some skill and it is better to use long lengths with no fittings at all.

Blue medium-density polyethylene is ideal for underground plumbing: it is not affected by chemicals in the soil and is not susceptible to pipe bursts. There are special 'compression' fittings you use for joining it to itself or joining it to copper: the size to use is 20mm.

The procedure for burying pipe underground is much the same as for laying drains – see Chapter 5. The trench needs to be deep enough for the pipe to be positioned below the depth that frost will penetrate into the ground and beyond the reach of normal digging – 750mm is the minimum.

If the tap is by a garden shed or greenhouse, take the opportunity of the trench to lay a garden electrical supply as well – in special armoured cable or normal cable laid in rigid plastic conduit.

Setting a post in concrete

Choose a stout post (preferably oak) at least 75mm by 75mm and about 1.5m long. Shape the top to an angle or fit a post cap so that it will shed rainwater. If the post hasn't been preservative-treated, soak it in wood preservative before you install it

Place a large stone at the bottom of the hole and surround the stone with gravel. Prop the post upright and pour concrete (1 part cement: 4 parts combined aggregate) round the post to a depth of about 0.3m, i.e. half-way down the length of the post in the ground. Slope the top of the concrete away from the post. When it dries the concrete will shrink back a little – fill the gap with creosote or wood preservative

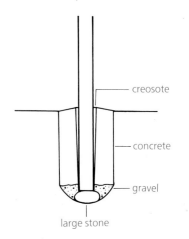

— creosote

— concrete

— gravel

large stone

Installing a tap down the garden

Use a bib tap and wallplate elbow for the garden tap

It isn't essential, but some sort of drain or water butt beneath the tap will stop puddles forming

Where the pipe begins to rise to the garden tap, it must again be lagged all the way up

Lay a continuous length of pipe underground with no fittings

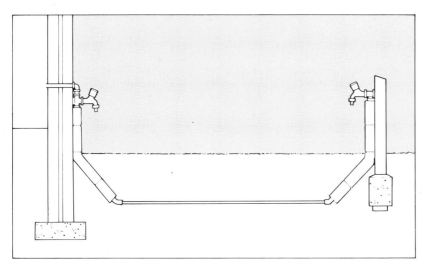

At the house end, you can take the supply from a tee off the branch pipe to the outside tap (outside the house) or from the wallplate tee fitting. This means that only one pipe has to pass through the wall and that both taps can be isolated by the same stopvalve

The pipe should be protected from frost by lagging from where it comes out through the house wall to a depth of 750mm below ground. Use pre-formed weatherproof pipe lagging and seal the top of the lagging with flexible mastic to stop water getting in

The rising pipe needs to be clipped to a well-fixed support. A wooden post bedded as deep as the pipe and held firm by concrete is best. A garden wall will also do – or a well-secured fence post, but don't rely on a fence panel

Installing outside taps

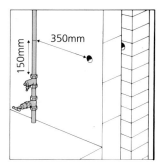

Choose a position for your tap where it will be convenient for connection to the rising main and for drainage

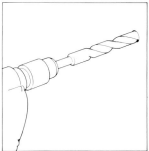

Make the hole through the wall by drilling with a long 16mm diameter masonry bit – if necessary, hire a drill and bit

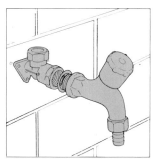

Drill and plug the wall and fix the wallplate elbow into which the tap screws. Insert a fibre washer and screw the tap into the elbow

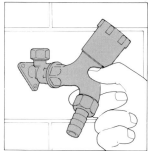

It's unlikely that the tap will sit upright when it is screwed fully home. Unscrew it and add washers until it sits correctly

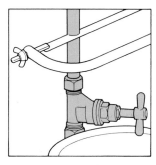

Turn off the rising main stopvalve and drain the water (draincock not shown). Make two cuts in the rising main and fit an equal tee

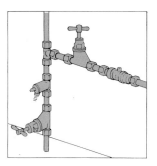

Plumb from the tee in 15mm pipe and fit a stopvalve (make sure the arrow is in the direction of flow) and a double-check valve

Continue in 15mm pipe, and by means of bends or elbows, take the pipe through the wall and connect it to the wallplate elbow

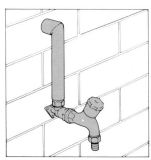

Check for leaks and tighten any joints as necessary. Protect the pipes outside with waterproof lagging. Drain down in winter

Making a manipulative fitting

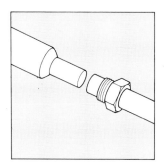

Prepare the ends of the pipes as for other fittings (see Chapter 2). The nut and joint body must be passed over the pipe first

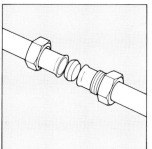

With the retaining nut in place, use the steel drift to shape the ends of the pipe. Turn the drift a little between hammer blows

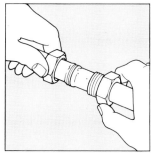

Smear jointing paste liberally over the cone and the two ends of pipe and assemble them together. Tighten the nuts

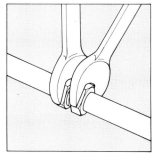

The completed joint is fully watertight and will not pull apart under the action of frost – which is why it is used outside

Ornamental waterfalls and fountains

A garden pond can be enlivened by the addition of a waterfall or fountain. There is usually no need for a permanent water supply to a pond; initially it can be filled by a hose pipe and thereafter rainfall will keep it topped up except in very dry summers. The water for the fountain or waterfall is pumped from the body of water in the pool.

There are two types of pump:

Submersible pumps sit in the water: they draw in water directly through a large strainer and pump it either straight to a fountain jet positioned on top of the pump or via a hose to a waterfall or fountain.

Surface pumps are positioned next to the pond; they draw water in through a suction hose and pump it out through another hose.

A pump's performance is measured in terms of the discharge it can maintain: the further it has to lift the water (i.e. pump uphill), the less the discharge will be: using narrow hoses will also reduce the output. In most pools, you will want to keep the flow down to the minimum required to circulate the water without causing too much disturbance to plants – water lilies in particular prefer still conditions and it's best to keep them at the opposite end of the pool to a waterfall. A discharge of 23 litres a minute (300 gallons an hour) is sufficient for a fountain 1.8m high. A fountain should be no higher than its distance from the side of the pool.

Electrical installation

Pumps need an electrical supply: because you are more vulnerable to shocks from electricity when you're working in the garden, it's safest if the supply is extra-low voltage – 24V or less. To reduce the house supply of 240 volts to 24 volts you need a **transformer**; pumps that run on extra-low voltage should come complete with one. The transformer has to be housed in the dry and the best answer (if it's practicable) is to put the transformer inside the house or the garage so that *all* the cable down the garden is at extra-low voltage. If you have to install the transformer some way down the garden, it will need a specially-built weatherproof enclosure.

If you're running a 240V supply down the garden, it is essential for it to be properly installed in the correct sort of cable, adequately protected and with a residual current device (sometimes called a residual current circuit breaker) fitted for safety. Even if you normally do your own electrical work about the house, this is a job for which you should consider employing a qualified electrician. Look for one who is an Approved

Contractor on the roll of the National Inspection Council for Electrical Installation Contracting (NICEIC).

If you're using a submersible pump, there will be a length of cable passing over the edge of the pool into the water. Protect this from accidental damage by passing it through a length of hosepipe and under flagstones, where possible. Don't hide an unprotected cable in foliage: there have been accidents to people pruning back foliage and cutting through a dark cable.

Plumbing a pump

With submersible pumps, mostly push-fit fittings are used. A fountain jet simply pushes on to the pump outlet. Tee-piece fountain jets have an outlet for a hose to a waterfall. The hose is usually held firm by a worm-drive ('Jubilee') clip.

With a surface pump, you need threaded fittings to connect the hose to the pump. It's best to include gatevalves in the pipe runs so that the pump can easily be taken out for maintenance and winter storage. Use reinforced hose for the suction side (to stop it collapsing) and fit a strainer to the end in the water. Install the pump in a proper housing with a drain at the bottom to allow water that collects to drain away.

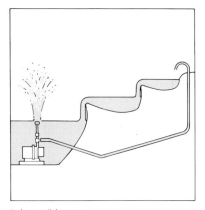

Submersible pump

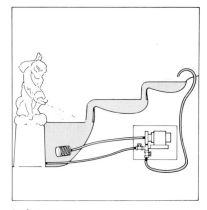

Surface pump

3

CENTRAL HEATING

About central heating

Installing central heating is one of the most valuable additions you can make to a house. Much of the work involved is no more difficult than other home plumbing work and you will be able to save a considerable amount of money by installing your own system. However, there are some parts of the job – connecting up a gas boiler, for example – where it is necessary to get qualified help. You may also need professional help with the design – which means choosing the best system for your household, selecting, sizing and positioning components such as radiators and boilers and deciding the layout of the system.

You can employ the services of a professional designer to make these decisions for you or you can enlist the help of one of the do-it-yourself superstores or mail-order companies who supply central heating equipment. Either way, you can carry out the actual installation work yourself; or employ a central heating contractor to do both the design and the installation.

Many central heating systems are installed to a standard pattern, often without much detailed planning or tailoring to individual requirements. Controls, in particular, are likely to be limited or old-fashioned and the system may not operate as efficiently as it could.

Proper design of heating systems takes into account both the practical considerations – such as fuel availability (and room for storage), space for equipment, house design and so on – and the requirements of the people living in the house.

Lifestyle will often determine the best choice of system and equipment and will certainly be paramount in deciding the best types of controls to have. If you choose the controls at an early stage, the design of the remainder of the system may well fall into place. For instance, if your house is occupied 24 hours a day, solid fuel is a possible choice; if it is occupied for only a part of this time, you would probably go for a fuel such as gas or oil which can be automatically controlled, turning the heat off completely when not required.

The way in which the heating system is designed will also depend upon whether the whole house needs heating at the same time, or whether the heating to individual rooms (or even a complete floor) could be turned down at certain times – maybe a room is occupied only at weekends, say. Equally, there may be a large room which requires only occasional heating at short notice. A fan convector type of heater would be an ideal choice here, giving out fast heat when required, but it would need to be on a two-pipe circuit – so this would immediately determine the best type of distribution system.

Choosing the central heating

Whether you are going to design and install a central heating system yourself or are going to get someone else to do the detail design and/or installation, there are some basic decisions you will have to make. An installer may make incorrect assumptions if he doesn't know what you want. The decisions you can make at this stage are:

□ the type of fuel to be used
□ how the system is to be operated and controlled
□ which type of boiler you want and where it is to be installed
□ which type of radiators or other heaters are needed in each room, and where they are to be located
□ what measures can you take to improve the insulation in your home and so reduce the heating load and therefore both installation and running costs.

Chapter 11 (starting overleaf) goes through the choices and looks in detail at the various components necessary for installation of a 'wet' central heating system.

An installer will be able to give you guidance on these points but in most cases you should be able to arrive at the best choice yourself.

Designing the system

If you are going to take the plunge and do the complete design and installation yourself (except for the gas and electrics), you will have to work through the following steps in addition to those above:

□ calculate the heating requirement for the house
□ determine the actual size (output) of the boiler and radiators
□ decide on the best way of running the pipework
□ size the pipework
□ size the pump
□ decide on the details and method of installation of all the other components.

Chapter 12 looks at each of these steps in detail; you should read right through to the end of the chapter before making any decisions. Although we have tried to present each stage in a logical sequence, there are inevitably a number of factors which are inter-related.

Doing the installation

After the theory comes the practice. Chapter 13 takes you step-by-step through the main stages of installing an open-vented 'wet' central heating system (see opposite). You will need to refer to the earlier chapters of the book, particularly Chapter 1 (*Tools*), Chapter 2 (*Pipes and fittings*), Chapter 3 (*The cold water system*) and Chapter 4 (*The hot water system*).

What is central heating?

A central heating system is one in which heat is produced at a central point and then distributed around the house. There are two basic types of system – 'wet' and 'dry'.

Wet systems

In this type of system, water is heated in a fuel-burning or electrically-heated **boiler** and is then circulated through pipework to **radiators** where the water gives up its heat and returns to the boiler to be re-heated. In modern systems, a **pump** is invariably used to circulate the water and a system of **controls** turns the heating on and off as required.

By far the majority of wet systems are the *open-vented* type: this has a **feed-and-expansion cistern** which keeps the system topped up with water and allows the water to expand into it as it heats up, plus a safety open vent pipe to allow steam to escape if the boiler thermostat fails. An alternative is the **sealed** system with a specially-designed

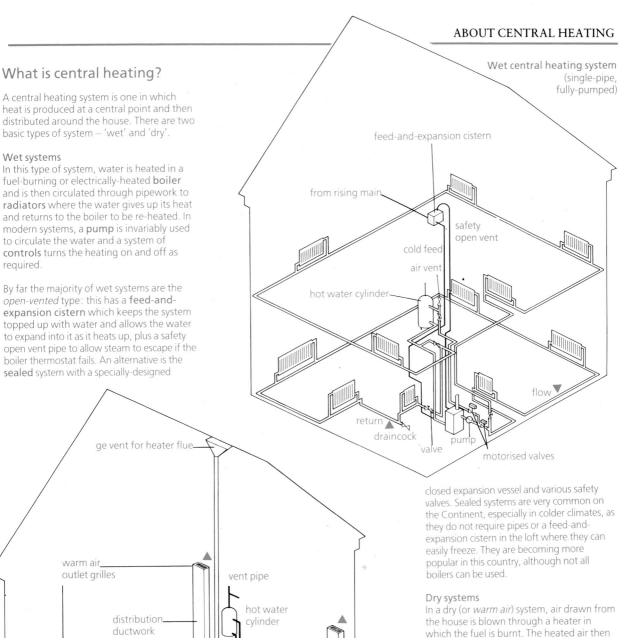

Wet central heating system
(single-pipe, fully-pumped)

feed-and-expansion cistern

from rising main

safety open vent

cold feed

air vent

hot water cylinder

return

draincock

valve

pump

motorised valves

flow

Dry central heating (including hot water heating)

ge vent for heater flue

warm air outlet grilles

vent pipe

distribution ductwork

hot water cylinder

heater

closed expansion vessel and various safety valves. Sealed systems are very common on the Continent, especially in colder climates, as they do not require pipes or a feed-and-expansion cistern in the loft where they can easily freeze. They are becoming more popular in this country, although not all boilers can be used.

Dry systems

In a dry (or *warm air*) system, air drawn from the house is blown through a heater in which the fuel is burnt. The heated air then passes through a series of insulated ducts, usually made of galvanised sheet metal, to outlet grilles which are usually located in the floor or just above the skirting board. The heated air then warms the room. To ensure good air flow round the house, ventilators are often fitted above doors. Before re-entering the heater, the air is filtered.

Most gas warm air heaters incorporate a small gas water heater which is connected to an indirect hot water cylinder.

Dry systems like this are usually restricted to new houses since the ductwork is difficult to install in an existing house.

CHAPTER ELEVEN

CHOOSING CENTRAL HEATING

The wide and potentially bewildering variety of components available for central heating systems, with a number of different manufacturers' products apparantly doing the same job, can make choosing a system something of a headache even once you have decided which fuel to use.

Which fuel?

Heat is usually produced in a boiler by burning gas, oil or solid fuel, or by direct heating from electricity.

Gas

Gas is the most popular fuel and supplies are available in most areas. Even if there is no supply to your house, it is quite likely that there will be a nearby main from which your Gas Board can run you a supply at a reasonable cost. Gas requires no storage facilities and burns very cleanly. Its other great advantage is that it can readily be turned on and off, so gas-fired boilers can be fitted with automatic controls.

Gas can also be used for gas fires in individual rooms or for gas convector heaters with a linked control system.

Oil

There are different grades of oil: the type used for domestic boilers is referred to as *gas oil* or '28sec' oil. The '28sec' is the oil's viscosity measurement. Oil requires considerable storage facilities – ideally at least 2700 litres – and whilst it does not burn quite so cleanly as gas, it is also suitable for automatic control.

Electricity

At the normal daytime tariff, electricity is extremely expensive, so for central heating it is used in conjunction with a thermal storage unit operating on Economy 7 tariff. This type of system is not always easy to control, particularly during the spring and autumn when the outside temperature can change over a couple of hours: a sudden rise in temperature could leave you with an unnecessary quantity of stored heat which will eventually have to be used up whether you need it or not. Whilst there are controls which vary the quantity of heat stored depending upon the outside and/or house temperature, a sudden *drop* in temperature could mean that you have to top up the heating using electricity at the very much more expensive daytime tariff.

Individual electric **storage heaters** also operate on Economy 7 and can provide whole house heating.

Solid fuel

You can get bituminous coal or a variety of smokeless types such as Coalite, Anthracite and Coalflow

Fuel costs compared	annual efficiency	Cost per kWh
Gas (conventional boiler)	70%	2.03p
Gas (condensing boiler)	85%	1.58p
Oil (28sec)	65%	1.92p
Electricity (general Tariff)	100%	6.05p
Electricity (Economy 7)	90%	2.47p
Electricity (90% at Economy 7 and 10% at day rate)	95%	2.98p
Solid fuel (anthracite grains)	60%	3.02p
Solid fuel (Coalflow pearls)	70%	2.78p
LPG (conventional boiler)	70%	3.72p
LPG (condensing boiler)	85%	2.91p

Pearls. Some boilers can burn bituminous coal in smokeless areas. You need storage facilities of at least 1.5 tonnes (a tonne of coal takes up around 1.3 cubic metres) and boilers and chimney or flue require regular cleaning. A solid fuel system is not generally suited to automatic control in the same way as a gas or oil system (though modern solid fuel boilers are better – see overleaf). Another disadvantage of solid fuel is that you will probably need an alternative method of heating the domestic hot water in the summer.

The burning of **wood** as an alternative to coal is now possible in suitably designed *dual-fuel* boilers provided you live outside a smokeless zone. You *must* keep the wood dry for at least a year before use, or there is a risk of serious damage to the boiler and chimney due to tarry deposits, which also present a fire hazard. It is important to ensure that the chimney is of suitable construction or you run the risk of structural damage. The Solid Fuel Advisory Service can give you advice on this and all other aspects of solid fuel use.

LPG

Liquefied petroleum gas has the same advantages as mains gas, but needs storage of at least 1200 litres.

Costs

The cost of fuels is constantly changing; the table opposite shows the approximate comparative cost per useful kWh output for the different fuels based on average prices in spring 1990. These take into account average annual efficiencies for a well-maintained system, correctly installed and with effective control of both the radiator and hot water heating circuits.

The maintenance and pump running cost will be similar for all fuels; standing/rental charges, which apply to electricity, gas and LPG, are excluded.

Choosing the boiler

The type of boiler you can have depends not only on the fuel to be used, but also on whether it is to be wall-mounted or floor-standing and whether it can use a conventional flue (chimney) or needs a room-sealed balanced flue.

Gas boilers are available in by far the widest selection of types and designs, both floor-standing and wall-mounted. Most manufacturers now offer conventionally-flued boilers, for connection to an existing lined chimney, and natural draught balanced-flue (or room-sealed) boilers which can, within certain limits, be located on or adjacent to any outside wall, through which the flue terminal is installed. Some balanced-flue models are **fan-assisted** which gives more flexibility over their position – see *Flues*.

There have been considerable improvements in the efficiency of modern gas boilers with the design of low water content heat exchangers, and efficiencies of up to 80% at full load can now be achieved. Wall-mounted boilers are available with coloured front panels which, together with their compact dimensions, means they blend well with fitted wall units.

A recent development is the **condensing** gas boiler which can have an efficiency exceeding 90% and, although more expensive to buy, will be cheaper to run. The increased efficiency is achieved by a larger heat exchanger which extends into the boiler flue outlet: the water returning from the radiators passes through the extension first and the heat in the flue gases is transferred to the water. The result of this is that the flue gas temperature drops below the point at which the water vapour in it will condense. The condensed water thus produced is piped away to a drain and the discharge of the lower temperature flue gases from the

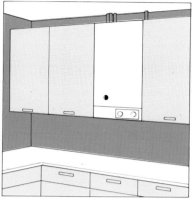

Wall-mounted, balanced-flue gas boiler

Condensing gas boiler

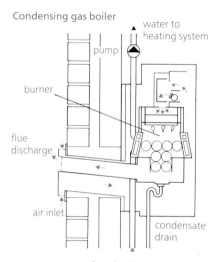

water to heating system
pump
burner
flue discharge
air inlet
condensate drain
water from heating system

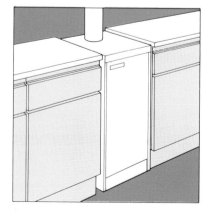

Floor-standing gas boiler with conventional flue (oil and solid fuel boilers are similar)

flue terminal is assisted by a fan.

Condensing boilers can only be fitted to fully-pumped systems and are not suitable for systems controlled wholly by thermostatic radiator valves (TRVs). Ideally, larger radiators are used with a lower water temperature.

Another new type of gas boiler is the **combination** boiler, which combines the functions of boiler and multi-point instantaneous water heater (see page 49). Its advantages are that you do not need a hot water cylinder or a feed-and-expansion cistern and that hot water is always available: the main disadvantages are that the hot water flow rate is slower and, in a hard water area, the boiler itself will suffer from scale unless some method of water treatment is adopted (see page 42).

Where normal gas supplies are not available, a number of gas boilers will also operate on **LPG**.

Oil boilers are also available in floor-standing or wall-mounted versions and with conventional or balanced flues. Most oil-fired boilers used to be of the **wallflame** type in which the oil ran in controlled quantities into a circular burner ring. Nowadays, these have given way to far more compact **pressure jet** burner units which are much quieter and less smelly than their predecessors.

At least one manufacturer is now making a **condensing** oil boiler.

Solid fuel boilers are all floor-standing and have to be filled with coal and emptied of ash, although this tiresome task has been reduced to a once-a-day operation with many units. All have to be used with a conventional flue.

Most modern solid fuel boilers incorporate **combustion air fans**, which can give a rapid increase in burning and hence a greater degree of control over heat output.

Some solid fuel boilers designed to burn Coalflow Pearls have underfeed

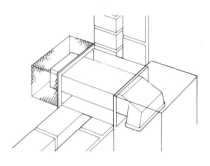

Balanced flue terminal

grates and automatic de-ashing and may need attending to only once or twice a week.

Electric boilers are extremely compact and require no flue and virtually no maintenance, but for domestic use on normal tariff would be very expensive to run. Units incorporating **thermal storage**, which charge up at the much cheaper Economy 7 rate, are very much larger and heavier than other boilers and would probably need to be positioned in a garage or outhouse.

Back boilers are available for use with both gas and solid fuel, but there can be a problem concealing unsightly pipework as it leaves the boiler. In addition, the position of fireplaces into which they might be installed is often remote from a convenient position for the hot water cylinder, which results in reduced efficiency in the summer. There is also the problem of locating the pump and controls in a convenient position remote from the boiler.

For **sealed systems**, the boiler (gas or oil) must be fitted with a high-level safety thermostat – either as original equipment or supplied as an add-on 'kit'.

Flues

Whichever fuel you use (apart from electricity), it will require an adequate supply of fresh air to ensure efficient combustion, and some way of getting rid of the products of combustion (which include water vapour).

Conventional flues If a chimney is inadequately insulated, the flue gases leaving the boiler will cool excessively as they pass through the chimney and water vapour will condense as droplets. With an unlined brick chimney, this water can be absorbed into the brickwork, leading to deterioration of the structure and possible staining on the inside walls. This can have serious consequences with oil-fired boilers, since oil also contains a small amount of sulphur which, when burnt, will be given off as sulphur dioxide, some of which forms sulphur trioxide. Normally this will remain as a gas and be discharged from the flue along with the other products of combustion. However, if condensation takes place in the flue, the sulphur trioxide will be absorbed by the condensed water to form dilute sulphuric acid. This will cause serious damage to the chimney structure and if it runs back into a steel boiler will corrode it.

It is therefore essential that there should be the minimum heat loss from the flue gases to avoid condensation, which will mean *insulating* the chimney, especially if it is against an outside wall. By far the most common and cheapest method for gas and oil boilers, where there is an existing chimney, is to install a flexible stainless steel **flue liner** which runs from the boiler flue pipe where it enters the chimney stack to a terminal at the top. The top of the stack is closed off by a blanking plate so that there is a sealed airspace between the outside of the flue liner and the stack – it is this airspace which provides the insulation. With oil-fired boilers, it is recommended that additional insulation is installed by filling the void around the liner with loose-fill insulation (such as vermiculite); it is essential that the liner is of the correct grade of stainless steel. If you have no suitable chimney, a purpose-made one can be installed.

A liner does not have to be installed for a solid fuel boiler if the chimney was erected before the 1966 Building Regulations came into force. For chimneys built after this, a liner must be installed. Flexible liners must not be used with solid fuel boilers, since they cannot be cleaned satisfactorily, they may not withstand the high temperatures that can occur if there is a chimney fire, and their slightly ribbed internal surface will readily collect soot deposits. For a straight chimney with no offsets, it can be lined by dropping down a

rigid **sectional liner** which can be of acid-resistant asbestos cement pipe, clay or refractory cement or double-skin stainless steel. Where the chimney is not straight, a specialist firm can line it using an inflatable 'sausage' of a suitable size inserted into the chimney with a lightweight insulating material pumped in around it. When this has set, the sausage is deflated and withdrawn, leaving a smooth lining.

If you do not have a convenient chimney, a completely new one could be built, either internally or externally, using twin-wall insulated stainless steel flue sections or purpose-made insulated concrete sections. The requirements for a new chimney are quite complicated and it is essential to contact both your local Building Control Officer and a firm of chimney specialists.

Balanced flues With gas and oil boilers, most of the flue problems can be easily overcome by installing a natural-draught or fan-assisted balanced-flue boiler; this also gets round the problem of providing a fresh air supply. The casing of this type of boiler is sealed from the room and a two-compartment duct connects from the boiler to a terminal on the outside wall. The flue gases leave the boiler through one of the ducts and fresh air passes into the boiler through the other. With the natural-draught type, the boiler has to be mounted directly on or adjacent to the outside wall and there are also certain restrictions as to the positioning of the terminal. With the fan-assisted type, some boilers can be fitted 2 metres or more from the terminal, the positioning of which does not have the same limitations.

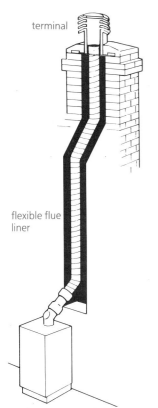

Flue lining for conventional chimney

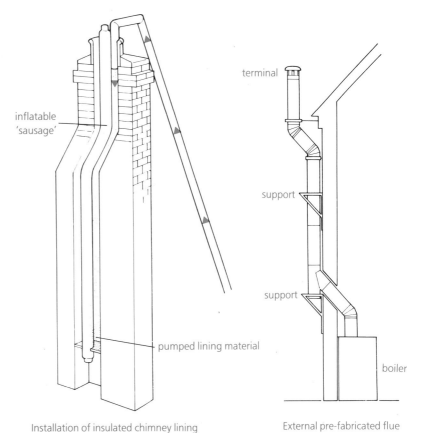

Installation of insulated chimney lining

External pre-fabricated flue

Siting the boiler

If you will be using solid fuel, your choice of boilers is confined to floor-standing units with a conventional flue. The boiler's location is therefore likely to be determined by the position of a suitable chimney.

If you have decided on a gas or oil boiler, you have the choice of floor- or wall-mounted models. Where you put it depends on whether you are going to use an existing chimney stack, install a new one or use a balanced flue. A balanced-flue boiler can be put in any room but a conventionally-flued boiler must not go in a bathroom or bedroom.

With any boiler, the pipe runs to the hot water cylinder should be kept as short as possible. You must also allow for running the fuel supply to the boiler and ensure that there is adequate access for installing and servicing the boiler. For an oil-fired boiler, you must take into account the position of the oil tank and the method of connection.

For a conventionally-flued boiler, there must be adequate fresh air provision. As a general guide, the minimum 'free' area of the fresh air inlet grille should be $550mm^2$ for every 1kW of boiler output over 5kW. If a gas boiler is in the same room as an extractor fan, this free area may need increasing to avoid the fan drawing flue gases out of the flue.

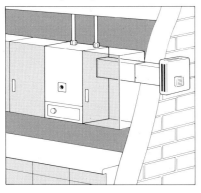

Fan-assisted, wall-mounted boiler can be positioned away from its balanced flue terminal

Choosing the radiators

Heat is transferred in three ways – conduction, convection and radiation. Conduction is where heat passes directly from one material to another or one part of a substance (solid, liquid or gas) to another (heat travels through walls by conduction); with convection, the heat is circulated by the movement of something like air with hotter air rising and cooler air falling. With radiation, the heat rays (from the sun or an open fire, for example) only heat things up when they meet them.

In order to feel comfortable, the air temperature around you *and* the average temperature of surrounding surfaces (including radiators) need to be warm, so that you receive heat from them, rather than give it out.

Panel radiators, despite their name, give out about 80% of their heat by convection and provide a good balance of convected and radiant heat for most circumstances. Where a room needs to be heated quickly, a fan-assisted convector-type heater may be more suitable. We have used the term 'radiator' generally to include panel radiators, convectors and skirting heaters.

Panel radiators

The steel panel radiator is the most common and cheapest type of heat 'emitter' for central heating. Both single and double panel types are available; some have banks of fins behind the panels. These (sometimes known as 'convector' radiators) give a considerably increased heat output without occupying any more wall space, which could prove very useful if you need to replace an existing radiator with one of higher output. In addition to the conventional steel panel radiator there is a wide range of high output radiators manufactured from steel, cast iron, aluminium and even plastic, in a variety of designs and colours.

Single panel and 'convector' radiator

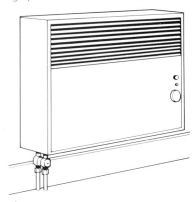

Wall-mounted fan convector

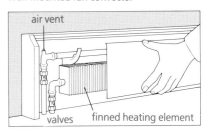

Skirting convector

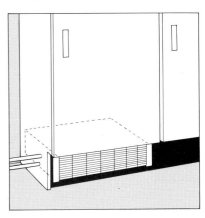

Plinth fan convector

Note that radiators in a one-pipe system (see below) should have the flow connection to the *top* of the radiator to achieve full output.

Convectors

As an alternative to a panel radiator, there is a variety of convectors, both natural-draught and fan-assisted types – the latter needs an electrical supply for the fan, but provides a faster heat-up. A convector **skirting heater** provides an effective way of getting an even distribution of heat into the room without being obtrusive. If you have suspended floors, there are several manufacturers of **underfloor convectors** which have combination inlet and discharge floor grilles in a variety of finishes. One manufacturer (Kampmann) produces an underfloor convector that can be fan-assisted for higher outputs and is shallow enough for installation within a 100mm joist depth. A compact '**plinth**' fan convector fits unobtrusively under fitted kitchen units, into a stair riser to heat a hallway, under bench seating or where restricted space does not allow the installation of a wall-mounted heater. It can be controlled remotely.

The pipework layout

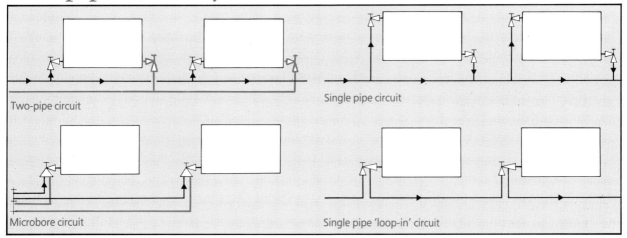

Two-pipe circuit

Single pipe circuit

Microbore circuit

Single pipe 'loop-in' circuit

To take the water from the boiler to the radiators, there are two main types of distribution system – one-pipe and two-pipe.

With a **two-pipe** system, which is by far the most common, two separate pipes, one a flow and the other a return, run from the boiler via a pump in a loop from radiator to radiator. Water passes into the radiator from the flow pipe and out into the return. This system has the advantage that the pump gives a positive flow of water through each radiator. It can use either **smallbore** pipe, which is mainly 15mm and 22mm and is also known as mini-bore, or **microbore** pipe, which is mostly 8mm, 10mm or 12mm. Where a microbore system is instal-led, the main flow and return pipes from the boiler, which will be in 15mm or 22mm pipe, connect to **manifolds** to which the smaller pipes running to and from the radiators are connected. Being flexible, micro-bore pipes can be threaded through floor joists in much the same way as electrical cables, which considerably simplifies installation and reduces costs, but the system has to be carefully designed; a more powerful pump may be needed to circulate the water through the small pipes.

With the **one-pipe** system (always smallbore), the water circulates

Microbore manifold

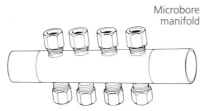

around one or more single pipe loops at each floor level, usually with the assistance of a pump, but the circula-tion into each radiator is by convec-tion with, ideally, the inlet at the top and the outlet at the bottom.

Although there is less pipework to install, the system must be carefully designed to make sure the radiators will work properly.

A slight variation on the basic single pipe system is the *loop-in* system, in which the pipe goes into a single two-connection valve on each radiator. The valve incorporates a by-pass so that some of the water enters the radiator whilst the remain-der goes through the by-pass and is joined by the cooler water leaving the radiator; the mixed water leaves the valve through the second connec-tion and passes to the next radiator.

Choosing the other components

Apart from the boiler and the radiators, there are many other components to choose for a central heating system.

Hot water cylinder

Details of the types of hot water cylinder are given in Chapter 4.

For operation with a central heating system, you need an **indirect cylinder** – that is one with an internal heating coil. You may be able to re-use or modify your existing cylinder, but if you are buying a new one, it is a good idea to go for the pre-insulated type. 'Packaged' cylinders come pre-plumbed and pre-wired with pump, motorised valve, thermostat, programmer and various other components: all you have to do is to connect them up.

Pumps

Modern domestic central heating pumps are extremely compact and the majority have adjustable speeds. Although they do not require any maintenance, they can stick after a period of non-use and, with their small internal clearances, any debris can cause them to jam. It is therefore worthwhile ensuring that you buy a pump that has a facility for rotating the impeller by hand, usually via a screwed access plug in the top. The majority of pumps can be bought complete with isolating valves which screw on to the pump inlet and outlet

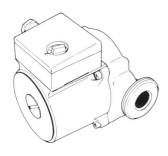

Circulating pump with variable speeds

connections and have compression fittings at the other end for direct connection to the pipework. If they are not provided, isolating valves should be bought separately.

Pipes and fittings

Despite the increasing availability of plastic pipe, the majority of central heating systems are still installed using copper pipework – see Chapter 2 for details. For joining central heating pipes, soldered capillary fittings are a better choice than compression fittings. Apart from being cheaper and less obtrusive, capillary fittings are much less likely to leak after installation. This is because the continual heating up and cooling down of the pipework, as the system goes on and off, will cause it to expand and contract slightly. This can result in a gradual and slight slackening of the tightened compression fitting olive, which can eventually leak, even if it was correctly installed in the first place. Often, the leak will seal itself after a time, leaving a greenish white deposit around the fitting, but the one that does not is bound to be under the floorboards!

Draincocks These are obtainable with threaded connections which can screw into a fitting with a female iron end or directly into a threaded connection on a boiler. It is worthwhile spending a little more on the type that has a sealing gland, otherwise water leaks out around the plug when it is open to drain and you have to use a dish underneath to catch the drips. Draincocks are also available built into stopvalves and tee or elbow fittings or with plain connectors which can be soldered directly into a copper fitting such as a tee (remove the plug with its rubber washer first), but these are not always of the gland type.

Air vents There are two types of air vent – manual ones that have to be operated by a key and automatic ones. Most radiators have *manual* air vents to release the air from them as they fill with water. Sometimes, these are supplied as integral fitments, though with many radiators they are supplied separately as threaded fittings (usually 1/2in BSP) which are screwed into one of the top connections. These threaded air vents can also be fitted into female threaded fittings to provide a vent point on the pipework. Also available are air vent fittings that can be soldered directly into a capillary fitting and ones that are integral with a fitting such as a tee or elbow.

There are two types of *automatic* air vent. One, called a 'hygroscopic' air vent, may still be found in many British homes. However, British Standard 5449 does not recommend this type of air vent because it may permit water to escape and so enable fresh water to enter from the feed and expansion cistern. The other type of automatic vent is the float type – the fitting contains a float which operates a small needle valve via a lever mechanism. Air entering the fitting escapes through the needle valve, but as the water reaches it the float rises so closing it off. Both these type of vents usually have 3/8in BSP threads so they are often used in conjunction with a 3/8in by 1/2in BSP reducing bush.

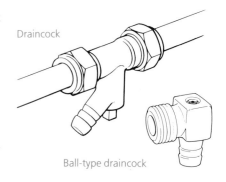

Draincock

Ball-type draincock

Hygroscopic air vent

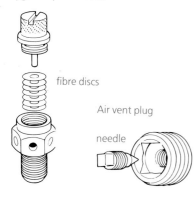

fibre discs

Air vent plug

needle

Ball-type valve

water flow

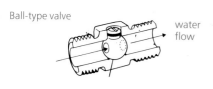

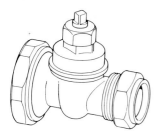

Pump isolating valve

Lockshield radiator valve with drain point

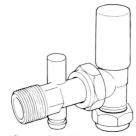

Wheelhead radiator valve

Isolating valves Gatevalves (see Chapter 2) can be fitted to isolate sections of the pipework system or components such as the hot water cylinder or pump. Ball-type isolating valves have a ball inside them with a hole through it. The ball can be turned by using a screwdriver or Allen key so that the hole is either in line with the inlet and outlet of the valve, allowing the water to flow, or rotated through 90 degrees so shutting it off. 'O'-ring seals prevent water leakage around the ball. This type of valve is extremely compact and is often used as means of isolation for a fan convector. They are commonly available in straight pattern 15mm sizes, usually with compression fittings at both ends but a wide range of types is available.

Radiator valves The valves fitted to radiators are available in both straight and angle patterns and with *wheelhead* or *lockshield* tops. The difference between these is that the wheelhead top will turn the spindle of the valve so that you can open or close it, whilst the lockshield top does not turn the spindle. Instead, the lockshield top can be removed so that a spanner (or a wheelhead top) can be used to adjust the valve and regulate the flow of water into the heater. This procedure is known as *balancing*. Once this has been done, the lockshield top is replaced and the valve setting cannot be changed accidentally. Radiator valves have a compression fitting at one end for connecting the pipework and at the other a threaded union male iron 'tail' for screwing into the radiator connection. For microbore systems, **twin entry** valves are available. They incorporate a long tube which carries the flow water into the far end of the radiator, preventing it mixing with the cooler return water leaving the radiator through the same valve.

See also **thermostatic radiator valves** (TRVs) on page 138.

Safety valve This type of valve is an essential requirement only on sealed systems, though one should always be fitted to an open-vented system if there is any possibility that either the cold feed pipe or the safety open vent pipe could freeze at any time. It has an internal mechanism much like a stopvalve, except that the jumper washer is held against the seating by a spring. If the pressure of water on the underside of the jumper washer exceeds that exerted by the spring, the jumper will lift so the water can escape through the valve. The spring tension can be adjusted but these valves are usually supplied pre-set and should not be altered. Many valves discharge through a series of holes in the top of the valve body. This could be dangerous if you are standing nearby since the water is likely to be above boiling point and turn to steam as soon as it escapes. A much more satisfactory type has a screwed outlet connection which enables discharging water or steam to be safely piped to a convenient adjacent position.

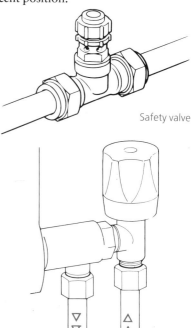

Safety valve

Twin-entry radiator valve

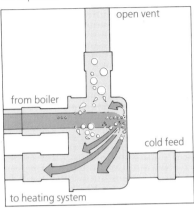

Air separator

open vent

from boiler

cold feed

to heating system

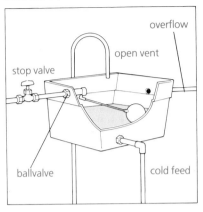

Feed-and-expansion cistern

overflow

open vent

stop valve

ballvalve

cold feed

With the trend towards low water content boilers, problems can be caused by small bubbles of air which continually circulate in the system, being carried straight past vent points and through radiators, causing noise at various places. This can be overcome by fitting an **air separator**: the water leaving the boiler passes through it and any bubbles of air collect and are bled off via the safety open vent pipe or via an automatic (float type) air vent.

Expansion cisterns and tanks

With an open-vented system, a **feed-and-expansion cistern** needs to be connected at the highest point in the installation. It can be made from galvanised steel, polypropylene or GRP and must be able to withstand boiling water. It will have a mains water supply, an isolating valve and ballvalve which will ensure that the heating system is kept full of water at all times. As the water in the system heats up it will expand in volume – by around 2.5 litres for the average system containing around 90 litres. This expansion takes place back up the cold feed pipe into the cistern, raising the water level. See Chapter 3 for details of fitting a cistern.

In the same way that the feed-and-expansion cistern accommodates the expansion of water in an open-vented system an **expansion vessel**

accommodates it in a sealed system. The tank is divided into two approximately equal sections by a rubber diaphragm – one section is connected to the heating system and the other is filled with air at a pressure of around 0.5 bar (7½lbs a square inch).

When the heating system is empty, the air pressure will push out the diaphragm into the water side. As the system fills, the water pressure will push it back until it is again in a central position. When the water heats up it expands back into the tank, pushing the diaphragm back into the air side; the pressure increases by around 1 to 1.5 bar for a correctly-sized tank.

With some boilers, the expansion vessel is accommodated within the boiler casing, together with the safety controls and pump. Otherwise, the tank is connected at a suitable point in the system: it measures approximately 280mm long with a diameter of around 234mm.

A sealed system is filled from the cold water main through a special fill connection. This is a water company approved fitting which incorporates a stop valve, double check valve and flexible hose. The hose must be removable and not permanently connected. A pressure gauge must also be installed in a prominent position and within view of the fill unit.

Improving insulation

It's sensible to improve insulation before installing central heating. Not only will this reduce running costs, but it will also mean that a smaller system can be installed.

Leaks through gaps are often responsible for very considerable heat losses so the first thing to do is to check all the windows and doors and remove any paint build-up around the edges or on the hinges so that they close firmly against the frames. Small gaps can be effectively sealed with draught excluders.

The next place to look, if you have a suspended floor, is at the floorboards and the gap below the skirting board. Any complete floor covering will reduce the air leaking into the room from below the house through gaps between the floorboards and a good quality carpet and underlay will provide insulation. Gaps at the bottom of the skirting can be sealed with silicone rubber sealant or the polyurethane foam available in aerosol cans. If you are installing a conventionally-flued boiler, you must maintain a permanent opening to outside so that fresh air for combustion can get in.

To reduce heat losses from the house itself, the obvious place to start is the loft space where you should have at least 100mm thickness of blanket-type insulation between the joists. Don't pack the insulation into the eaves – you need to keep the loft space over the insulation ventilated to prevent condensation. Don't forget to insulate and draughtproof the loft hatch.

The next stage is to consider insulating the walls: with cavity wall insulation if they are of cavity construction or by lining internally with insulation-backed plasterboard if they are solid. Double glazing, although improving comfort and allowing radiators to be placed elsewhere than under windows, does not save a large amount of heat.

Choosing controls

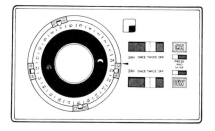

Traditional programmer

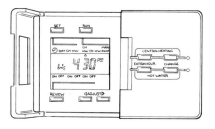

Modern central heating programmer

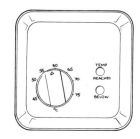

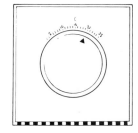

Room thermostats

The way in which most control systems work is quite simple. The basic principle is that you have a **programmer** which you connect to a mains electrical supply and set to different times. When it reaches the time at which you have set the heating and/or hot water to come on, internal switches operate and an electrical supply is sent to the heating and/or hot water **thermostat**. If a thermostat is calling for heating (i.e. the temperature is too low), an electrical supply is passed on to, say, a **motorised valve**. This operates to allow the water from the boiler to flow to the heating or hot water cylinder circuit as required, and at the same time the boiler and the pump are switched on. When the thermostat is satisfied (i.e. the temperature is high enough), it breaks the circuit and the boiler and pump switch off.

The principle is the same with a **controller** except that the thermostats are replaced by **detectors** and the controller operates the valves, boiler and pump directly.

Programmers and controllers

These are an integral part of any heating system since they enable both heating and hot water to be automatically switched on at pre-set times. Modern electronic **programmers** are much improved from the older electro-mechanical types, being silent in operation and offering greater flexibility. They often give up to six timed periods a day and a battery reserve, so that you do not lose the programmes if the electrical supply fails. Many can be programmed separately for every day of the week with independent programmes for hot water and central heating.

A **controller** is more sophisticated than a programmer since as well as giving timed control of the system, it is connected to detectors to provide switching based on temperature. Most can be separately programmed for each day of the week and some have the facility for controlling at different temperatures at different times of the day. The use of detectors gives much closer temperature con-

trol than thermostats, with resultant fuel saving.

A problem can arise where thermostatic radiator valves (see overleaf) are fitted to all the radiators: they may all close down when the rooms are up to temperature, but the boiler and pump will continue running, with the boiler wastefully cycling on and off. With most of the valves closing down, the reduced water flow may be below the minimum permissible level for the boiler. This can be overcome by fitting a permanent by-pass between the outlet side of the pump and the return pipe to the boiler, though this will reduce the operating efficiency of the system. An alternative is to fit a pressure-controlled valve in the by-pass, which opens when most of the thermostatic radiator valves are closing, but this arrangement will not stop the pump and boiler continuing to operate after the valves have shut down. And the way of solving *this* problem is to fit a **flow controller** which consists of a flow sensor located in the pipework and connected to a controller. When the flow drops below a pre-set adjustable level, as a result of the thermostatic valves closing down, the boiler and pump are stopped, so saving fuel. Every so often the pump is started again – if the flow has increased, due to valves opening again, it remains running and the boiler is switched on; if not, it is switched off again.

The most sophisticated form of control for central heating systems is a **boiler manager** – either a compensator or an optimiser (or both).

A **compensator** senses the outside temperature and the boiler water temperature and uses this information to compute the necessary boiler firing and pump running time. An **optimiser** allows you to set the actual times you need certain temperatures and it then 'learns' when to turn the system on to achieve these.

Thermostats and detectors

A **thermostat** is simply a switch which operates automatically in response to a change in temperature. It contains a sensing element, which moves slightly as the temperature changes. This movement operates an electrical switch. An adjustment knob enables the temperature at which the switch operates to be varied. For a **room thermostat**, this is typically 3°C to 27°C (37°F to 81°F); for a **hot water cylinder thermostat**, usually the strap-on type, it is 40°C to 90°C (104°F to 194°F).

A point to bear in mind with thermostats is not to worry too much about the actual temperature reading they are set to – what is important is that the temperature in the house is as you want it. Remember, too, that during colder weather, when the walls of the house will be colder, you may have to raise the thermostat setting to maintain the same comfort level.

The function of the **boiler thermostat** is to turn the boiler on and off to keep the water leaving the boiler from rising above the set temperature. The thermostats are sometimes calibrated in numbers rather than temperatures and some are just switches marked 'high' and 'low', but they all have maximum water temperature settings of around 82°C to 85°C.

A **frost thermostat** is sometimes installed *outside* the house to bring on the heating if the temperature falls below freezing. With a house heated regularly, this is normally unnecessary since it would take the inside temperature (which is what matters) a long time to drop to freezing level. The only occasion when this could matter would be if the house was being left unoccupied for a while – in which case the inside thermostat or thermostatic valves could simply be turned to a low setting. If the boiler or some of the pipework is in a completely unheated area, the frost thermostat can bring on just the pump to keep the water circulating and, if it is considered really necessary, a separate pipe thermostat can be connected to bring on the boiler if the water temperature drops to near freezing.

A **detector** also senses the temperature of its surroundings and sends a signal to a separate electrical controller which carries out the electrical switching and where the required temperatures are set. Detectors work at low voltages, usually 12V or less, from a transformer in the controller. Under no circumstances must they be connected to a mains supply.

Valves

Central heating systems can have **motorised valves** which are opened or closed by a small electric motor rather than by hand. The valves, which usually have compression pipe connections, are available as either two-port or three-port types. A two-port valve is either open or closed, but with a three-port valve, the water enters one port and leaves by either or both the other ports, depending upon the type of valve and the needs of the system. Many motors incorporate transfer or auxiliary switches which can turn the boiler and pump on and off. Some motors require an electrical supply both to open and to close them; others require a supply only to open them – they close automatically under the action of a spring; this can considerably simplify the wiring.

A **thermostatic valve** consists of a valve body and an operating head containing a fluid or vapour-filled bellows. As the temperature changes, the bellows expand or contract and this movement opens or closes the valve. The most common type is the **thermostatic radiator valve** (TRV) which can be adjusted to keep a room at any temperature you select, typically between 5°C and 25°C: as the room temperature approaches the set point (adjusted by rotating the valve head), the expansion of the bellows gradually shuts off the flow of water into the radiator. They are available in both angled and straight patterns, with twin entry to suit microbore systems, and as low resistance valves for single-pipe systems. They can also have *remote sensors* for situations where the valve head is in a position not representative of the room temperature (e.g. concealed by furnishings) or could be affected by heat from pipework or the sun. For other applications, thermostatic valves are available for use as zone control valves or control valves for the hot water heating.

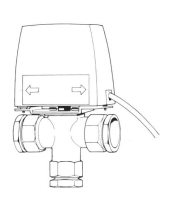

Three-port motorised valve (a two-port valve is similar without the bottom connection)

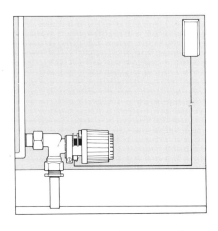

Angled thermostatic radiator valve with remote sensor

Examples of control systems

The two examples on this page assume that the system heats both radiators and hot water.

System A This system uses a programmer, a three-port motorised control valve, a room thermostat and a cylinder thermostat. An important consideration is that the *room thermostat* will control heating throughout the house so it must be installed in a representative position and away from draughts – often the first-floor landing is a good location. The *motorised valve* operates in conjunction with the boiler and pump and under the control of the *programmer* and thermostats. It allows the water to circulate from the boiler to the house heating circuit, the hot water cylinder heating circuit or both, as required.

A controller could be used instead of the programmer with detectors in place of the thermostats. You could also use two separate two-port motorised valves, one on each of the heating circuits, in place of the three-port valve (see Box). However, with a low water content boiler, where the pump is often wired via the boiler control circuit so that it runs for short periods after the boiler has been switched off, using the three-port valve (which will always be open to one circuit) will avoid having to incorporate a boiler by-pass.

The location of the room thermostat is nearly always a compromise and, wherever you put it, some rooms will end up too warm – perhaps because the sun is shining through a window or several people are in the room – while the thermostat located elsewhere is still calling for heat. It is here that the *thermostatic radiator valves* come in useful since they can be fitted to selected radiators or convectors to give control in individual rooms. However, they must not be installed on radiators serving the area in which the room thermostat is located. You may

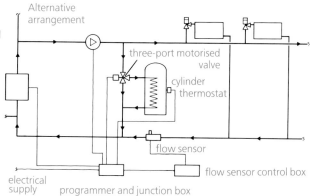

System A

junction box

room thermostat

electrical supply

programmer

radiator

two-port motorised valves

three-port motorised valve

by-pass (if required)

hot water cylinder

cylinder thermostat

System B

flow sensor control box

electrical supply

programmer

thermostatic radiator valves

pump

sensor

boiler

two-port thermostatic valve

flow sensor

Alternative arrangement

three-port motorised valve

cylinder thermostat

flow sensor

flow sensor control box

electrical supply

programmer and junction box

also have a room (perhaps a bedroom) which is not always in use so you can save heat by fitting a thermostatic valve in that room, enabling you to keep it at a lower temperature when it is unoccupied – or, of course, you can turn off an ordinary radiator valve.

With a larger house, where several rooms are unoccupied for periods, you could divide the house heating circuit into two *zones*, one for the main rooms and one for the other rooms. Both zones would be fed via the three-port valve but the second circuit would also have its own

two-port control valve. This valve would be controlled by either a manual switch or a separate time switch so that it remained closed whilst the heating to the remainder of the house was on. You could additionally have a thermostat in one of the rooms on the second zone to control the two-port valve.

System B This system has thermostatic valves on all the radiators and a two-port *thermostatic valve* with a *remote sensor* controlling the hot water cylinder heating, the sensor being strapped to the side of the cylinder. A *flow sensor* is installed on the common heating return back to the boiler so that the boiler and pump switch off if all the thermostatic valves are nearly closed. A programmer controls the times during which the boiler and pump operate.

If the boiler is the type requiring a minimum flow of water to be maintained, a three-port thermostatic valve (plus a cylinder thermostat) would be used to control the cylinder heating. When the stored hot water is up to the required temperature, the valve operates to divert the heating water to by-pass the cylinder and flow straight back to a connection to the return pipe (after the flow sensor connection). The thermostat shuts off the boiler if its operation is not required. The flow sensor will still sense the flow in the heating circuit.

Although this system gives you individual temperature control over each room, it does not allow the heating and hot water to be run at different times – this could mean that if you don't use much hot water during the day, the cylinder is being unnecessarily kept up to temperature all the time. In this case, it would be better to replace the three-port thermostatic valve by a three-port motorised valve and cylinder thermostat, but connected in the same way. This would enable the hot water heating to be separately controlled: see the *Alternative arrangement*.

CHAPTER TWELVE

DESIGNING CENTRAL HEATING

Designing a complete central heating system properly is a time-consuming operation. So is installing it, and installation also causes disruption throughout the house. For most people, the least complicated way is to employ a firm to design and carry out the complete installation. Usually, but not always, the same firm does both jobs. This method is the most expensive, of course, but if you choose your firm carefully you should get a good system. Even if you decide on this course, much of the information in the rest of this section will still be useful in helping you to discuss your needs and choices sensibly.

The alternative is to do the work yourself – either the whole job or just the installation following a design drawn up by someone else.

Doing the installation yourself and letting someone else do the design work is a good compromise. If the design work is accurate and comprehensive, the job is not difficult.

Getting a system designed

There are two main ways of getting a system designed – using a firm selling equipment, or employing a professional heating engineer.

Some firms specialise in selling central heating equipment, mainly by mail order. A few of these also offer a design service for a relatively nominal fee, which may be refunded if you buy the components from the firm. The firm is unlikely to see your house: instead you draw a scale plan of it and fill in a short questionnaire. The design the firm produces cannot be better than this questionnaire and your replies. To get the best deal, make sure you give the firm any extra information you think could be relevant and tell them any special needs. Read through the whole of this chapter, so that you are aware of the assumptions the firm will probably make and will be able to foresee possible pitfalls.

Most of these firms will provide a custom-made design, with 'isometric' drawings showing the pipe runs and wiring diagrams, and you will be able to telephone the engineer who did the design with any queries.

If you have a particular requirement – such as a boiler manager or a condensing boiler – a firm will be able to build this into the design for you. Many firms sell helpful booklets about central heating.

Many of the d-i-y superstores who sell central heating components offer a central heating design service for free (or very cheaply), but often the design you get will be a variant of a standard system, using well-tried and tested components, with just the number and size of the radiators (and, perhaps, the size of the boiler) altered from one design to the next, together with standardised advice on installation. They also sell booklets on central heating.

Some heating engineers specialise in designing heating systems for pri-

vate homes. This sort of personal service is bound to cost more, but the fees could be a good investment if the resulting system is well designed and economical.

A heating engineer will not only choose and size all the components for you, but he will also suggest where they should be installed and how all the various pipes should be run – so all you have to do is to install them and connect them up.

Don't forget manufacturers' leaflets and brochures which are often a good, and usually a free, source of information on design and installation. Whoever sells you your components should give you the manufacturer's full instructions for the major parts – boiler, pump, controls – but as most of these instructions also contain design information, they are often worth sending off for.

More technical help is available in literature and booklets published by the different fuel authorities.

Doing your own design

Designing central heating from scratch means first of all finding out how much heat the system needs to produce to keep the rooms at the temperature you want. This is not difficult but it is tedious as each room has to be considered in turn.

Armed with the information about the heat requirements, you will be able to choose the size of boiler, radiators (or convectors), pipes and pump you need, and to plan how they are all to be positioned and connected.

Calculating the heating load

To calculate the amount of heat that you need to put into a house, you actually work out how much would be *leaving* it with each room at its 'design' temperature and a certain outside temperature. Heat is lost through the different parts of the building structure and through gaps.

The rate at which heat flows through the building depends on:

□ the materials and method of construction of different parts of the house
□ the thickness and area of the materials used
□ the temperature difference between the inside and outside of the building.

Different parts of the house lose heat at different rates – depending on whether the materials used are good or bad *insulators* and the way in which the different parts of the house (walls, roof, etc.) have been constructed. The factor which describes this property for any part of the structure is called the thermal transmittance or **U-value**. The rate of heat loss, in watts, is then given by the U-value multiplied by the area through which the heat is passing and the temperature difference between inside and outside.

---

rate of heat loss
equals
$U \times A \times (T_i - T_o)$

where:

U = U-value

A = area in square metres

T_i = inside temperature in degrees Celsius

T_o = outside temperature in degrees Celsius

The units of U-value are watts per square metre per degree Celsius (W/m²°C). But as this description is a little cumbersome, it is often omitted, and books tend to give U-values simply as a bare number. This can lead to confusion with older publications, where the U-value may be given in its old imperial unit which is an even more cumbersome Btu/h/ft²°F. (To convert from imperial to metric units, multiply the imperial figure by 5.7.)

U-values for the different parts of a house are given in **Table 1** overleaf.

To calculate the actual heat loss you have to decide on what the inside and outside temperatures are going to be. The recommended levels for each room are: living rooms (including dining rooms, bedsitting rooms and so on) 21°C; bedrooms 18°C; kitchens, halls etc 16°C; bathrooms 22°C.

Remember these are *maximum* design temperatures – it is of course possible to run the heating system so that rooms are cooler if you want them to be.

The external temperature used in this calculation should be the lowest it is likely to get for any length of time: minus 1°C (−1°C) is the figure usually taken. For walls between rooms, the 'external' temperature is, of course, that of the room on the other side of the wall. If this is close to the temperature of the room you are considering, the heat gain or loss is normally ignored.

However, heat loss into adjacent rooms (or rooms above or below) should be considered where there may sometimes be large temperature differences – a living room will lose heat into a bedroom left unheated above it, for example, or into an unheated house next door.

Ventilation loss

The amount of heat lost through cracks or gaps around doors, windows and so on varies considerably from house to house and it is not possible to calculate it accurately. The heat lost depends on how much air moves out of the house. Many designers use a very basic rule of thumb and estimate the number of times the volume of air in the house changes every hour, assuming an average level of draught-proofing:

□ living rooms (including bedsitting rooms) at one air change per hour
□ bedrooms at half an air change
□ kitchens and bathrooms at two air changes
□ halls and stairways at one and a half air changes.

For rooms with open fireplaces and chimneys, *add* one air change.

The calculation means working out the volume of the room (length × width × height) in cubic metres and multiplying this by the temperature difference and 0.33 to give the number of watts required.

---

ventilation heat loss
equals
$n \times V \times (T_i - T_o) \times 0.33$

Where:

n = number of air changes

V = room volume (in m³)

T_i = inside temperature (°C)

T_o = outside temperature (°C)

Table 1 U values for normal exposure

	W/m² °C
walls – external	
solid brick, 105mm thick plastered inside	3.0
with 22mm insulation	1.37
with 28mm insulation	1.19
solid brick, 220mm thick plastered inside	2.10
with 22mm insulation	1.15
with 28mm insulation	1.02
cavity, brick outside and inside with plaster	1.50
with cavity wall insulation	0.55
cavity, brick outside/block inside (lightweight)	0.92
with cavity wall insulation	0.46
walls – internal	
brick, 135mm including plaster	1.8
lightweight block, 100mm plus plaster	1.1
concrete block, 100mm plus plaster	2.2
timber stud and plasterboard covered	1.6
doors	
wood, in internal walls 32mm	2.1
in external walls 45mm	2.0
fully glazed, in internal walls	3.6
in external walls	4.7
floors – intermediate	
timber, heat flow upwards	1.6
heat flow downwards	1.4
floors – ground [1]	varies from 1·5 to 0·45
windows	
single glazed metal	6.4
single glazed wood	5.0
double glazed wood (and plastic)	2.9
double glazed metal, with thermal break or aluminium/PVC composite	3.7
double glazed metal, without thermal break	4.3
roofs	
pitched, without felt	2.7
pitched, with felt	2.6
pitched 25mm insulation	0.99
50mm insulation	0.61
80mm insulation	0.42
100mm insulation	0.35
125mm insulation	0.28
160mm insulation	0.23
flat uninsulated	1.66
with 50mm insulation	0.54
with 100mm insulation	0.32

[1]. For ground floors, the U value depends on the size of the floor and the number of exposed edges. As examples:

	2m × 2m	4m × 4m
suspended	1.27	0.96
solid 1 exposed edge	0.74	0.45
solid 2 exposed edges	1.22	0.72

Heat Loss Chart Room: Bedroom 1

Surface	Size	Area	Temp-diff °C	U-value W/m²°C	Heat loss	Notes
Front window	1.8 × 1.22	2.2	19	5.0	209	All windows single glazed
Front wall (less front window)	3.4 × 2.5 (−2.2)	6.3	19	0.92	110	All external walls cavity brick/block
Side window	0.92 × 0.64	0.6	19	5.0	57	
Side wall (less side window)	4.2 × 2.5 (−0.6)	9.9	19	0.92	173	
Bed 2 wall	3.8 × 2.5	9.5	0	1.1	0	
Bed 3 wall	3.0 × 2.5	7.5	−3	1.1	−25	(Normally ignored)
Landing door	2.0 × 0.76	1.5	2	2.1	6	
Landing wall (less door)	1.6 × 2.5 (−1.5)	2.5	2	1.1	6	
Floor	3.4 × 4.2 (plus door recess)	15.3	−3	1.6	−73	Downstairs is warmer so heat flow is upwards (Normally ignored)
Ceiling	3.4 × 4.2 (plus door recess)	15.3	19	0.35	102	

Total surface heat loss 565W

Ventilation loss
(½ an airchange) 120W

Total heat loss **685W**

Calculating heat loss

To help you understand the calculations, here is an example, using a typical three-bedroom house the plan of which is shown right. The first step is to measure the area of each part of each room — we are going to take Bedroom 1.

When working out heat loss, do the calculations for windows and doors before the walls in which they are set. For example, the front wall has a window with an area of 2.2 square metres, a U-value of 5 and the temperature difference between inside and out is 18−(−1) = 19°C. So the heat loss through the window is:

$5 \times 2.2 \times 19 = 209W$.

To calculate the area of the front wall, subtract the area of the window and then do its heat loss sum. The wall has an area of 6.3 m² (8.5 m² less 2.2). Assuming cavity wall construction (U-value of 0.92), the heat loss will be:

$0.92 \times 6.3 \times 19 = 110W$.

The other outside wall is treated in the same way.

Note that, because the adjacent bedroom (a bedsitting-room) is 3°C warmer, there is a slight heat gain (25W) through one wall: this can be ignored. The same goes for the lounge below which is 3°C warmer than the bedroom (73W gain) and the landing which is 2°C cooler (12W loss). The other adjacent bedroom is at the same temperature, so there is neither heat gain nor loss.

Finally, there is the heat loss through the ceiling. This is 15.3 sq m and, assuming 100mm of loft insulation (U-value of 0.35), the heat loss will be:

$0.35 \times 15.3 \times 19 = 102W$.

The ventilation loss can be calculated assuming half an air-change an hour and the loss is:

$0.5 \times 38.2 \times 19 \times 0.33 = 120W$.

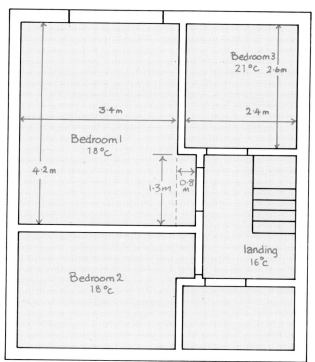

First floor

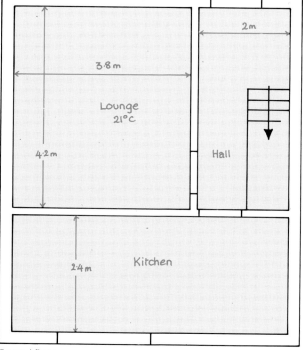

Ground floor

All the heat losses can be entered on your Chart (see left) and the procedure repeated for each room in turn to give the total heat loss for the house.

Notes
Outside temperature = −1°C
Ceiling height 2·5m
External walls — cavity brick & block
Windows — single-glazed, wood frames
Ground floor — solid
Roof — pitched with 100mm loft insulation
Internal walls — 100mm block

Sizing the boiler and radiators

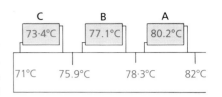

C	B	A
73·4°C	77.1°C	80.2°C

71°C	75.9°C	78·3°C	82°C

Calculating the average radiator temperatures in a one-pipe system

If the boiler and radiators were sized to match precisely the calculated heat load, it would take a considerable time for the house to heat up from cold. Also, it would not allow for the fact that the heat loss calculation will not give a precise figure – a number of variables relating to the structure itself (e.g. moisture content) can affect the heat loss – so the calculated figure can only be assumed to be accurate to within about 10%.

The normal practice is to add between 10% and 15% to the calculated heat loss to obtain the radiator size – the lower figure is used if the house is continuously heated. It is not usual to consider heat gain from people and equipment. However, if your system design results in a lot of exposed pipework in the rooms, the output of this can be deducted from the required radiator output. The following reduction can be made for each metre run:

15mm:40W, 22mm:56W, 28mm:72W

Sizing radiators

To select the size of radiators (or convectors), it is first of all necessary to know the system flow and return water temperatures. The normal temperatures are 82°C flow and 71°C return giving a mean water temperature of 76.5°C for radiators on a two-pipe system. If you are installing a one-pipe system, the mean water temperature of each heater will reduce as you go around the system.

The single pipe loop (above) has three radiators on it, A, B and C, with outputs of 1.5kW, 1.0kW and 2.0kW (total 4.5kW). The flow temperature at the start of the loop is 82°C and at the end 71°C. The temperature drop over the whole loop is divided between each radiator in proportion to its output.

So for A, the temperature drop is a third (1.5 ÷ 4.5) of 11°C, which is 3.67°C. The temperature of the water after the radiator will be 78.3°C (82°C – 3.7°C) and the *mean* (or average) temperature in the radiator will be 80.2°C.

Repeating the calculation for B and C gives mean temperatures of 77.1°C and 73.4°C.

Having arrived at the mean temperatures, you will then have to refer to the manufacturer's literature which will provide figures showing the heat output for each type and size of heater. For radiators, these are usually based on a room temperature of 20°C and the water flowing into the radiators at a temperature of about 82°C. If you are designing with different room or water temperatures, you will have to apply a correction factor – these will be listed in the catalogues.

There may be several different radiators with heat outputs near the one you want. But despite the hundreds of sizes available, it is quite likely that there will not be one of exactly the output calculated. The usual choice is to select the next size up, on the grounds that this will certainly be big enough. Going one size up is a very good idea in a living room, but for other rooms, if the next size down is only slightly too small (say one or two per cent) then go for this – your calculations are unlikely to be accurate to more than a few per cent and the smaller radiator will be a little cheaper.

If a room is large then it may be advisable to use more than one radiator in order to give a more even heat distribution – as a general guide, one radiator will satisfactorily heat an area of up to 20 square metres. Make sure that the height of the radiator you select will fit in – there should be a least 100mm clearance from the top of the radiator to any sill above it and 150mm clearance below for the pipework and connections. A long, low radiator will give better heat distribution than a tall one.

There are several factors that can reduce the output of a radiator and these must be taken into account when selecting a suitable size:
□ painting with a metallic paint will

For an average semi-detached house with solid walls (1930s, say), typical radiator sizes might be:			
Living room	(1 outside wall)	600mm × 1350mm	(double panel)
Hall/landing		600mm × 1400mm	(double panel)
Bedroom	(1 outside wall)	600mm × 1250mm	(single panel)
Bedroom	(2 outside walls)	600mm × 800mm	(double panel)
Small bedroom	(2 outside walls)	600mm × 900mm	(single panel)
Bathroom	(2 outside walls)	600mm × 800mm	(double panel)

reduce the output by between 10% and 25%, but this can be reduced if it is finished with two coats of clear varnish.

☐ a radiator shelf over it will reduce output by 10% (but will protect decorations or deflect heat outwards)
☐ putting a radiator in an open recess will reduce output by 10%
☐ encasing it with a grilled front will reduce output by 20% or more.

Sizing the boiler

The boiler has to be big enough to cope with the load of all the radiators and the pipework plus an allowance for heating the domestic hot water. It is important, however, that it is not oversized because most boilers are less efficient when running at part load. The boiler output should be based on the calculated heat load for the whole house plus a margin of 10% for heat loss from pipes and fittings (insulated pipes will still lose heat). Add a further margin of 10% (15% for solid fuel boilers) if the heating is intermittent.

After calculating the boiler capacity necessary for the heating circuits, the hot water heating has to be considered. With the margins already allowed, it is usually unnecessary to add anything for hot water heating, especially if a high-recovery cylinder is used with a control system which gives priority to the hot water.

If, after all these sums, the calculated output is borderline between two boiler sizes, but just creeps into the higher size, consider ways of reducing the necessary boiler output by around ten per cent – by adding more insulation, for instance. The cost of doing this should be offset by buying and running a smaller boiler.

Typical boiler sizes are: 6 to 8kW for a terraced house, 11 to 14kW for a three-bedroom semi-detached house and 16 to 18kW for a four-bedroom detached house, but good insulation could halve these figures.

The pipework layout

At this stage, it is a good idea (if you have not already done so) to draw up a plan of each floor so that you can mark the heater positions and piping layout. An *isometric* drawing is useful for locations where pipework is congested.

First mark on your plan the position of the boiler and hot water cylinder and then decide on the radiator positions. The normal place to position a heater is under the window since the hot air rising from it will counteract the effect of the cold down-draughts from the inner face of the glass. You must ensure that the bottoms of any curtains come down only to the sill level above the radiator. The sill then acts as a deflector, preventing the heat rising up and being trapped wastefully between the curtains and glass; a small shelf achieves the same effect.

If the window is double glazed, or in a bedroom, this is not so important and it is often possible to reduce pipe runs by putting the radiators back to back on internal walls. Make sure that you avoid obstructing any electrical socket outlets. If you are using a fan convector, its position does not matter so much, provided the path of the airflow is not obstructed by any furniture, but it must be in a convenient position for an electrical supply.

Having sorted out all the radiator positions, you can now plot the pipe routes – these should be kept as short and direct as possible. If you are putting a radiator or perhaps a heated towel rail in the bathroom, it is a good idea to connect it from the outlet side of the pump before any control valves. It will then heat up whenever the boiler comes on and not be dependent on the remainder of the house requiring heating. If you are installing a solid fuel boiler, you will need to connect this radiator on a separate gravity circuit from the

Radiator positioned under window

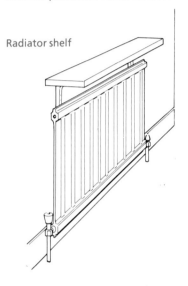

Radiator shelf

boiler to provide a *heat sink* as required by the boiler manufacturer. This will give the boiler's output somewhere to go whilst it is 'idling', ie when neither the house nor the hot water is calling for heat.

If you find that you can position most of the heaters on one or two pipe loops, you may consider installing a one-pipe system. You will have to re-size the radiators to take account of the reducing water temperature along the pipe and the radiators themselves must be connected top and bottom at opposite ends and always with the return connection downstream of the flow. Fan convectors will not operate satisfactorily on a one-pipe circuit.

It is worth experimenting with different pipe layouts, perhaps using a microbore system. If you are going to do the installation yourself, it is a good idea to inspect the floorboards at this stage to see which way the joists run and check whether there are any underfloor obstructions, such as a steel beam at first-floor level which spans the opening cut for a through-room. If a radiator is positioned parallel with the joists, check that there is not a joist directly below that will obstruct the pipes dropping under the floor from it – you may have to run surface pipework to the radiator or re-position it.

Solid ground floors can cause problems, too. Unless you're prepared to dig channels (and to put in accessible ducts), the pipes will have to be run on the surface.

Table 2

Provisional pipe sizes

total radiator output W	provisional pipe sizes mm
less than 650	6
650 to 1300	8
1300 to 2100	10
2100 to 5500	15
5500 to 11,000	22
11,000 to 18,000	28
18,000 to 28,000	35

Table 3

Actual pipe sizing

pipe size mm	heat loss uninsulated W/m	insulated W/m (thickness)	total load possible W
6	20	4 (9mm)	700
8	27	6 (9mm)	1,500
10	32	7 (9mm)	2,500
15	46	10 (19mm)	6,000
22	63	12 (19mm)	13,000
28	78	14 (19mm)	23,000
35	95	15 (25mm)	34,000

Sizing the pipes and the pump

Before the design is complete, you have to decide the size of the pipes in the system and choose the correct circulation pump (though most systems can use a standard-size pump).

Sizing the pipes

The larger the pipe, the more water (and thus heat) it can carry: each section of pipe must be large enough to carry enough water to provide the heat for *all* the heaters that are connected to that pipe – including all those that branch off it. This means the pipes nearest the boiler should have the largest diameter. Depending on your design, these pipes might also have to carry the hot water circuit load; pipes nearest a single radiator have the smallest diameter: usually 15mm in a smallbore system, 8mm in a microbore one.

Table 2 lists *minimum* pipe sizes for different radiator loads based on an 11°C temperature drop. Using this table, you can experiment with plans for different pipe layouts. Altering which radiators are fed off which circuit, or adding and removing circuits, can have an effect on the pipe sizes.

Normally, with a smallbore circuit, most runs will be in 15mm pipe, with perhaps just one long main run in 22mm, and only a small amount of 28mm pipe: note that the table gives 'provisional' pipe sizes. This is because the pipes have to be large enough to provide not only the heat for the heater, but also the heat that is lost from the pipes themselves. The amount of this loss depends on the length of piping, on whether it is insulated or not, and on the pipe diameter – and until you have designed a layout these are unknown factors.

The provisional pipe sizes make a generous allowance for heat loss from pipes, so it is unlikely that your pipe sizes will turn out to be too small, but it could happen if you are on the borderline between one pipe size and the next. It may be that the allowances are too generous, leading you to use a bigger pipe size than necessary. So after the provisional design, go back over the layout and check pipe sizes, this time calculating the actual pipe heat losses and checking that the pipe is the correct size for the actual total load (pipe losses and radiator outputs) it is to carry.

Table 3 gives the heat losses a metre from different sizes of pipe, and the maximum load that each can carry, based on an 11°C temperature drop and good insulation.

Sizing the pump

The pump has to be capable of pushing water round the circuits fast enough so that the water can give out the right amount of heat: the faster the water circulates, the higher the *average* water temperature and so the more heat is given off.

The speed at which a pump can push the water depends not only on the size of pump, but on the resistance the water meets in its travels – this resistance is greatest in small pipes, in long pipe runs and in runs with lots of pipe fittings or bends. Resistance also increases as the speed of the water through pipes increases. Modern pumps (see Chapter 11 for details) can cope with high resistances and still push water fast enough to cope with most normal heating loads. So it is not usually necessary to go through the calculations for pump sizing – just as well because they are complicated.

However, if your house is large, or your layout unusual, and if you intend to use a lot of microbore piping (which has a particularly high resistance), it is essential to go through the sums – it would be best to get a professional designer to do this for you.

Finalising the design

The feed-and-expansion cistern necessary for an open system should be at least 1m above the top of the highest heater or section of circulating pipework – this could be difficult if you are heating a loft conversion.

The feed-and-expansion cistern will also need a fitted cover; its capacity will normally be 45 litres. The cold feed and safety open vent pipes should both be 22mm and, if possible, connected to the heating flow and return pipes at the boiler to avoid pumping water into the cistern or drawing air in at the safety open vent. Sometimes the boiler manufacturer's literature will show an alternative arrangement that you can follow. There must be *no* restriction , such as a valve or pump, between the boiler and the termination of the safety open vent pipe.

Wherever possible, the pump should be put in the flow pipe from the boiler (after the safety open vent pipe connection), rather than in the return pipe, to avoid the possibility of air being drawn into the system at valve glands, joints, etc.

If you are going to install an air separator (which is advisable, especially with wall-mounted boilers) this should go on the flow from the boiler with the pump after it.

Now is the time to decide where you need isolating valves, draincocks and air vents. **Isolating valves** are usually installed to allow the pump, control valves and hot water cylinder to be removed without having to drain the system. **Draincocks** will be required at all the low points of the system to enable it to be drained, and also at the boiler – a drain point is often provided as standard on floor-standing boilers. **Air vents** will be

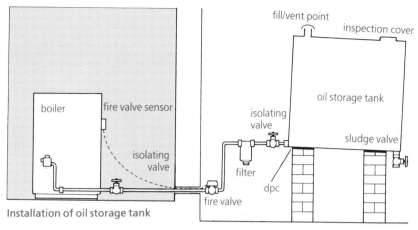

Installation of oil storage tank

required wherever there is a section of pipe which forms a high point from which the air cannot escape (e.g. through a radiator) when the system is filled. It is a good idea to mark all these fittings on your layout drawing and also to draw a *schematic* layout which shows all the main items of equipment. Do not forget the boiler by-pass connection if required by the boiler manufacturer for your layout.

Now check your layout to follow the path the water will take with the control valves in their various positions. This is to make sure that the water cannot circulate in reverse through any part of the system, by-passing the boiler or giving heat in unwanted areas. If this can occur, try to modify your piping or else fit a check ('non-return') valve.

If you are installing an LPG or oil boiler, you will need a storage tank and connecting pipework. With LPG, your supplier will be able to advise you of the requirements and the tank will be provided on hire. It is generally a requirement that the tank be visible from the position of the filling lorry. For an oil system,

you will need a purpose-made tank, which will be mounted on supports (with a 'fall' of 20mm in the metre) within a bund wall. The oil level in the tank must at all time be above the level of the boiler burner as the oil flows by gravity, and since the tank is a potential fire and pollution hazard you must notify your local Building Control Officer that you want to install one. He will advise you of necessary safety and constructional requirements for the tank supports, siting and enclosure. The tank needs to be fitted with a minimum 50mm vent pipe with protected outlet, a minimum 32mm fill pipe with cap and chain, a plugged drain valve and a contents gauge. The oil supply pipe to the boiler needs to be fitted with an isolating valve, filter and fire valve (connected to sensor). Your fuel oil supplier should be able to give you advice on the installation and the necessary pipe sizing and routing.

Oil tanks should be protected with a bituminous-type paint and there should be a dpc under the tank to prevent corrosion.

CHAPTER THIRTEEN

INSTALLING CENTRAL HEATING

Now you have reached the interesting part – carrying out the actual installation and seeing whether all your calculations have turned out right.

Most of the problems with installing a central heating system are likely to be over the siting of the boiler and the layout of the piping in the boiler area – so the instructions given here concentrate on those problems which are part design/part installation. It is important always to follow the manufacturer's instructions, especially where they differ from those given below.

If you need technical help or information while installing a system, ask the firm who sold you the components, the system's designer (if you got this part of the job done professionally) or the manufacturers. One job for which you *will* want a qualified person is making the fuel connections and checking the installation of a gas-fired (and perhaps an oil-fired) boiler.

Installing the boiler

There are standards to be followed for the installation of some boilers if Building Regulations are to be met. These mainly concern solid fuel boil-

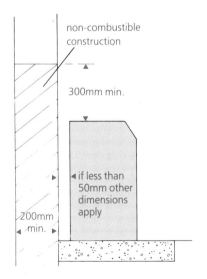

Positioning of solid fuel boiler

ers – domestic gas and oil boilers are usually designed so that they can be mounted on any wall or floor. Solid fuel boilers have to stand on a solid, non-combustible hearth at least 125mm thick and 840mm square. Usually a boiler will be within about 150mm of a wall at the back or sides; the wall has to be non-combustible and at least 75mm thick. If the boiler is closer than 50mm, the non-combustible part of the wall needs to be 200mm thick and to extend 300mm above the boiler (see diagram). Different regulations apply for solid fuel appliances in fireplace recesses (e.g. room heaters): if the fireplace already exists, its construction will probably be approved.

Before installing the boiler, read the manufacturer's instructions, including any references to permanent fresh air inlet requirements. After carefully marking out the position of a wall-mounted boiler, or placing a floor-standing one in position, the first job to tackle is installing the flue. The diagram shows the basic requirements for the positioning of balanced flue terminals.

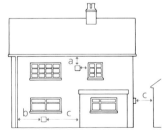

Location of balanced flue

Terminal position:	min
Below openable window or air vent	300mm
Below balcony (a)	600mm
Below gutter or eaves (a)	300mm
Above ground	300mm
From projecting pipes (b)	75mm
From corners or facing walls (c)	600mm

(In the solid fuel boiler diagram: non-combustible construction; 300mm min.; if less than 50mm other dimensions apply; 200mm min.)

Installing the boiler flue

A **balanced flue** is generally much easier to install than the conventional type, but even so it is important to get it right. The easiest way to cut the hole through the wall is first to drill holes from the inside around the marked outline and then to work (with hammer and bolster chisel) from both sides. This reduces damage to the wall surfaces.

The standard balanced flue terminal supplied with most boilers is telescopic and slides to take up wall thicknesses in the range 230mm (9in) to 380mm (15in). This will suit most walls but for particularly thick house walls or the one-brick-thick wall of an outhouse or garage you will need to specify the size.

To fit a balanced flue, follow the manufacturer's instructions, which usually call for sliding the two pieces to the correct position for your wall thickness and taping the joint between sections. Then position the flue in the wall and hold it in place with the internal backplate – this is screwed to wallplugs inserted in drilled holes in the wall. The boiler flue outlet seals on to this plate. The wall is then made good inside and out, effectively sealing the cavity.

A **conventional** flue can sometimes pose problems so check what size you need and whether you need specialist advice or assistance or have to notify your local authority. Have the chimney swept first.

The boiler connections to a prefabricated chimney vary from make to make: check with the manufacturer of your chimney system.

With an existing chimney, first fit a length (at least 600mm) of 'Urastone' flue pipe or sheet steel (usually with a white vitreous enamel finish) vertically to the boiler outlet, and a 135 degree bend with a soot door on top. All joints between flue pipes (except where connecting to a liner) have the socket facing upwards, are packed with caulking string, and

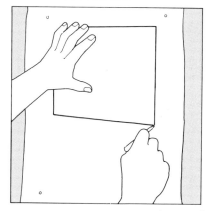

Marking the position of the balanced flue on the wall

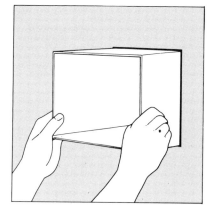

Fitting the terminal sleeve into the hole through the wall

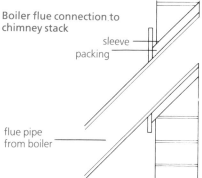

Boiler flue connection to chimney stack

sleeve
packing
flue pipe from boiler

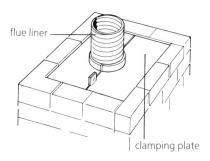

Fixing the top of a flexible flue liner

flue liner
clamping plate

sealed with a thick layer of fire cement. Break into the chimney at the point the bend meets the wall, forming a hole at roughly the angle the pipe will enter. Cement a metal or asbestos-cement sleeve (about 15mm bigger in diameter than the flue pipe) into the hole and then pass the flue pipe into this, caulking and packing the joint. If you are not using a flexible liner, cut the flue pipe so that it is flush with the inner wall surface of the chimney.

If you are using a **flexible metal liner**, lower this down the chimney from the roof after removing the chimney pot. If you don't want to carry the liner up on to the roof, it is possible (and safer) to pull the liner up from the bottom of the flue to the top with a rope, if there is little resistance and the flue is not too long. Push the liner into the socket end of the flue pipe. This can be

made easier by adding another 135 degree bend to the flue pipe which enables the liner to be dropped into it, but a larger hole has to be cut in the wall to make the joint. In some cases, for instance when the chimney is outside and therefore cold, you will have to provide a condensate drain at the lower end of the liner, but this is unusual. A warm flue is much better.

If you are contemplating any work that involves going up on to a roof, the greatest care should be taken to avoid any accidents. Do not attempt roof work in poor weather conditions, i.e. when it's wet and windy. The Health and Safety Executive produce two booklets – Safety in roof work (HS(G)33) and The safe use of ladders, stepladders and trestles (GS31) – which give information on the proper safeguards and are available from HSE or HMSO.

A clamp plate holds the top of the liner to the chimney stack – mortar this in place, then secure a terminal on top. Make sure that the flue pipe from the boiler is adequately supported and, if the draught diverter cannot be separated from the boiler, install a maintenance clip or 'split-collar' in the flue from the boiler. This is used to join two plain ends of flue pipe so that they can be separated and the boiler moved without disturbing the connection into the chimney stack. If required, insulate the space around the flue liner by pouring in loose-fill insulation from the top. Seal the gap between flue and collar.

Connecting the boiler

Most boilers have threaded tappings for connecting the pipes; frequently, these are 1in BSP, which may be larger than you need. They can be reduced by using a black malleable iron reducing bush but if the top tapping – the flow – is horizontal, by reducing the outlet to ¾in BSP you will trap a pocket of air at the top of the heat exchanger. You should therefore connect from the boiler with the same (or equivalent) size pipe as the tapping (e.g. 28mm for 1in tapping, 22mm for ¾in tapping) and reduce in size only at a point where the air can escape (e.g. after an upward-turning bend) or where you can provide a vent point on the pipe. The connections to the boiler should be made with copper to male iron capillary union couplings rather than iron to copper compression couplings – they are better able to withstand stress that can sometimes occur if the boiler settles very slightly on its fixings. It is also better to use jointing compound and hemp rather than PTFE tape on 1in threads.

You may be able to connect an oil-fired boiler to the tank at this stage: a gas-fired boiler, will be connected up later – see *Connecting gas* on page 153.

Installing the radiators

Hang the radiators on the wall before running the pipes but after checking that an underfloor pipe route is not obstructed. If it is necessary, paint their backs and bottom edges first. Radiator brackets have both circular holes and slots for fixing. After checking the position to give the required clearance from the floor and any sill above, drill and plug the wall to fix the first bracket, screwing through a slot and a hole so that it cannot slide up or down. Fix the second bracket at first with one screw through the centre of its slot. Then carefully put the radiator on the brackets and check its level – it should have a *slight* rise towards the end at which the air vent is situated.

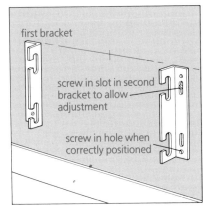

first bracket

screw in slot in second bracket to allow adjustment

screw in hole when correctly positioned

Fixing radiator brackets

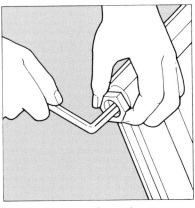

Screwing in radiator plug or air vent

After making any necessary adjustments to the position of this second bracket, carefully remove the radiator without moving the bracket and put a second screw through one of the holes to secure it.

Many radiators have four tapping positions – with a two-pipe system, you connect the radiator valves to the bottom tappings, the air vent to one of the top ones and a sealing plug in the other. The plug and air vent will be supplied with the radiator, often loosely screwed into the tappings. Remove them and re-fit at the correct ends after wrapping several layers of PTFE tape around the thread. You will require a special square-ended wrench to screw in the air vent holder and probably the plug as well. Fit the air vent needle and tighten it immediately so that it does not get lost.

The valves are fitted in exactly the same way after first separating the tail of the valve from the body. Depending on the make of valve, you may need a special radiator spanner for tightening the tails. If the same make of valve is at each end, the bodies will be interchangeable.

Conventionally, the wheelhead valve goes on the flow end. However, if you are fitting thermostatic radiator valves you must check the

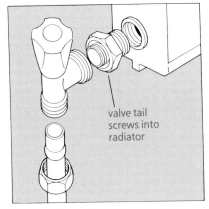

valve tail screws into radiator

Angled radiator valve

Reducing bush   Capillary union coupling

Running the pipework

direction of water flow which will be indicated on the valve body: some valves have to go on the return or with the heads horizontal.

With a single-pipe system, the only variation is that the flow should connect to the top of the radiator, so the plug will be in the bottom tapping at the same end. If you are using twin-entry valves for a single-pipe loop-in or a microbore system, you will need an additional sealing plug.

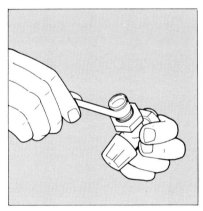

Wrapping PTFE tape on to valve thread

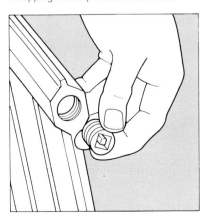

Air vent plug

Wherever possible, run pipes parallel to joists and supported on a piece of wood fixed between them. They can be clipped to the sides of the joists provided their route is clearly marked on the floorboards – someone may hammer a nail into the joist and miss. Avoid electric cables.

If you have to run across a joist, ensure that the notches are no more than 3mm deeper than the pipe diameter. As a general guide the notches should not be deeper than 0.15 times the joist depth, otherwise you may weaken it. Always place a piece of thin insulation between the pipe and the joist to minimise expansion and contraction noise. With a microbore system, you can thread the pipes from the manifold to the radiators through holes drilled in the centre of the joists.

The pipes must be supported at the following distances:

	horizontal	vertical
15mm	1.4m	2.4m
22mm	1.4m	2.4m
28mm	1.7m	3.0m

It is a good idea to use the plastic pipe clips which completely encircle the pipe – they prevent the pipe jumping out when it expands. With microbore pipes, keep to the same supporting distances as for 15mm pipe to avoid excessive sagging.

Never position a pipe in contact with another pipe or any part of the building structure. Always leave adequate room for fitting insulation (if necessary) and for expansion of the pipe – copper pipe will increase in length by approximately 1.2mm per metre at its maximum temperature. If you have ended a length of pipe with an elbow which is in contact with some part of the building – a wall or joist, for instance – the pipe will bend when it expands.

Pay particular attention to the level of the pipes – try to raise them very slightly towards connections to radiators so that any air in the pipe will pass into the radiator where it can be released via the air vent. If air is trapped at a high point in the pipework, you cannot rely on water pushing it out and you may have to install an air vent. With microbore pipe, the higher water velocity should clear the air.

Where a pipe passes through a wall, drill a neat hole that will just give clearance for it. If this is not possible, slip a sleeve or a piece of pipe, one size larger, over it before making the hole good. The sleeve should be of the same length as the wall thickness. The purpose of this is to prevent any mortar or plaster used in making good coming into contact with the pipe (this can cause corrosion) and to stop cracking of plaster around the pipe due to expansion movement.

When installing long runs of pipe under the ground floor, it is a good idea to slip lengths of insulation over them as you go, but do not cover any joints – you will need to check them for leaks. A self-grip ('Mole') wrench can be used to hold the insulation out of the way while you make a joint. The joints should be insulated after the system has been checked.

At least one drainvalve should be fitted at the lowest point of the system, and where a hosepipe can be easily fitted to drain the contents out of doors. In order to flush a system adequately – such as on completing the installation – the drain point needs to be a 15mm branch pipe fitted with a gatevalve and a hose connector. This will give a much better flow than a normal draincock. Down-loops of pipes – such as downstairs radiators fed from upstairs circuits – need their own drainvalves, as does the boiler.

Pipes laid across joists

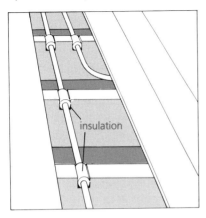

insulation

Air vent fitted to cylinder heating coil

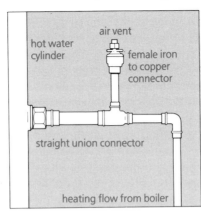

hot water cylinder

air vent

female iron to copper connector

straight union connector

heating flow from boiler

Feed & expansion cistern

Small cisterns are generally rigid plastic, but make sure the one you buy is suitable for use as a feed-and-expansion cistern, where it may have to withstand boiling (or near boiling) water. Connect it up in much the same way as a normal cold water storage cistern (see Chapter 3): fit a ballvalve through one side near the top supplied by a 15mm pipe run from the rising main (*not* from the storage cistern), a (plastic) 22mm overflow pipe and a 22mm feed pipe about 50mm up from the bottom of the cistern.

In order to meet the water bye-laws, you must use a diaphragm type of ballvalve with a servicing valve on its supply pipe. Adjust the float so that there is only about 100mm of water above the feed pipe connection when the cistern is first filled: the level will rise as the water in the system heats up. Secure the cover to make it near airtight and thus reduce water loss by evaporation and discourage the entry of bacteria. Insulate the cistern and the pipes, especially if they are in the loft.

The cistern should preferably be at least 1m above the highest point of the heating system – usually the top of the radiators or the heating coil in the hot water cylinder.

The safety open vent pipe should rise above the cistern and then turn down, passing through a hole in the lid, to terminate just above the overflow connection level.

With a **sealed** central heating system, you will not need a feed-and-expansion cistern – nor cold feed and open vent pipes. Instead, you will have an expansion vessel and a mains fill unit. The manufacturers will provide details of suitable points in the system where these components can be connected.

Cold feed and open vent

The cold feed from the feed-and-expansion cistern should be connected to the return pipe as close as possible to the boiler. There must be no valves in this pipe. Although normally a 15mm pipe, 22mm would be far better. This reduces the amount of corrosive oxygenated water that is drawn from the cistern into the system every time the water contracts as it cools down.

Connect the safety open vent, which should also be 22mm pipe, as close to the boiler as possible on the main system flow and *before* the pump. The pipe from the connection point to the boiler must be no smaller than the vent pipe size. Remember that you must *not* install any valves between the boiler and the cistern and that the safety open vent must terminate over the feed-and-expansion cistern and **not** the main cold water cisterns.

Some boilers have four tappings – two for a pumped circuit and two for a gravity circuit. With a fully-pumped circuit, you will need only two so the other two can be reduced down to connect to the cold feed and open vent unless the instructions specify otherwise.

Installing the pump

Buy a pump fitted with valves, for ease of servicing. These are jointed to the pipes with compression fittings and the pump slides in between them. Before installing the pump, check the instructions on how to mount it – the motor shaft may need to be horizontal. It is a good idea to mount the pump in a vertical sec-

Circulating pump

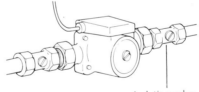

isolating valve

tion of pipe so that any trapped air vents via the pipework – for mounting horizontally, the pump may have an air release point. Have at least 400mm of straight pipe either side of the pump to reduce noise. After you have checked for fit, remove the pump and valves, and replace with a piece of plain pipe (using compression fittings). Leave this pipe in place until the system has been filled and flushed out. Then replace the pump.

Hot water cylinder

Unless you are going to fit a cylinder conversion unit to an existing direct cylinder (see Chapter 4 for details), you will probably be installing a new indirect cylinder. If you have old galvanised iron or lead pipework in the hot or cold water system, this might be a good time to consider replacing it – Chapter 4 also explains how to do this.

As with the boiler, it is preferable to connect to the cylinder with capillary union couplings. For the primary heating flow and return connections, you can get straight union connectors which have cone joints specifically for this job. If you are running a smaller size flow pipe (usually 22mm) than the size of the connection to the cylinder (often 1in or 28mm), increase the pipe size *before* it reaches the cylinder so that you can install an air vent point. Otherwise, air will be trapped in the top of the cylinder heating coil, reducing its efficiency and perhaps causing a noise of running water.

For the cold feed and domestic hot water connections to the cylinder, straight male union connectors with parallel threads should be used. If, as is often the case, the cold feed connection will end up behind the cylinder in an inaccessible position, connect it with a straight copper to female iron capillary connector before putting the cylinder in position. From this, run a short stub of pipe into one of the 22mm connections of a reducing tee. The second 22mm connection is joined to the start of the cold feed pipe and the reduced connection to a length of 15mm pipe which is brought forward to an accessible position and connected to a drain valve. The cold feed pipe is run to a compression fitting above the top of the cylinder, so as to be accessible for connection to the remainder of the cold feed pipe once the cylinder is

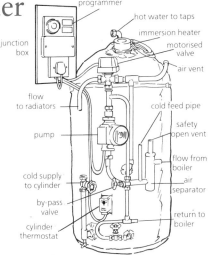

'Packaged' hot water cylinder

programmer
hot water to taps
immersion heater
junction box
motorised valve
air vent
flow to radiators
cold feed pipe
pump
safety open vent
flow from boiler
cold supply to cylinder
air separator
by-pass valve
return to boiler
cylinder thermostat

in position. Life will be a lot simpler if you go for a **packaged** cylinder, where all the connections are at the front.

Plan your pipework layout carefully so that it does not obstruct the access to the cylinder – you may have to remove it again if there is a problem with it (very occasionally a cylinder can leak at a seam) or if one of your pipe connections leaks. You may have to fit valves for the primary heating connections positioned to one side of the cylinder so that the pipes from the valves to the cylinder can be disconnected and removed in order to move the cylinder.

Connecting the gas

For the running and connecting of the gas supply, you need to call in a registered gas installer who will also test the boiler. It is therefore a good idea to leave the gas (but not water) connection until the rest of the system is complete and tested, so that only one visit is necessary.

Installing the controls

The requirements here will vary considerably, depending on the equipment you are going to use. Most control manufacturers provide details showing how their own controls are installed and connected, but it is not always easy to decide how items from different manufacturers should be interconnected electrically. This makes it sensible to use a control 'pack' from a single manufacturer, provided it has the selection of controls you want.

Unless you are competent to carry out electrical work in accordance with the Wiring Regulations, it would be advisable to call in a qualified electrician who is familiar with control wiring (not all are). The following notes are therefore confined to the physical installation of the control equipment and not the electrical connections.

Cylinder thermostat or detector

This is usually held to the side of the

Cylinder thermostat

hot water cylinder by means of a metal band – the best position for a single thermostat or detector is a third or so of the way up the side from the base. The alternative is a thermostatic valve with a remote sensor strapped to the cylinder.

Note that the insulation on pre-insulated cylinders needs cutting away to ensure good contact.

Programmer

This is usually wall-mounted near the boiler or the hot water cylinder. It doesn't really matter where, it just needs to be convenient. A programmer may need a mounting box.

Junction box

This is included to make the wiring up easier: all the cables from the various points in the system meet here. The junction box needs to go where it can be worked on easily and, to keep wiring runs short, near to the main components (which probably means close to the boiler, pump and motorised valve). The electricity supply cable is run from a fused switched connection unit fitted with a 3A fuse.

Cables

Wiring diagrams will give details of the types of cable needed from the junction box to each component – these may be twin (two-core), three-core, four-core or even five-core, all with earth. All cables are at 240V (except the low-voltage wire leading to detectors), but carry little current. Normal (PVC-insulated) wiring cable can be used except where heat-resistant ones are called for.

Room thermostat or detector

You may need to use a plaster-depth (16mm) mounting box for this.

Experimenting with position is worthwhile (but you must take care that any temporary wiring used for doing this is absolutely safe). In any event, the thermostat should be placed 1.5 metres or so above the floor, on an inside wall where air can flow freely round it and away from draughts and heat sources (including television sets).

Motorised valves

The position of these depends mainly on the pipe layout. Check whether the orientation of the valve and its motor is important, and if so, run the connecting pipes so you can attach the valve the right way up.

Room thermostat

Three-port motorised valve

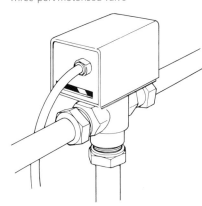

Final steps

Filling the system

Open all the radiator valves, but close all drainvalves and radiator air vents. Open the isolating valve to the feed-and-expansion cistern and let the system fill – this may take some time. Check continually for leaks. Those at compression fittings can usually be stopped by gently tightening the nuts. With solvent-weld plastic and capillary fittings, you have to drain the water out of the system first – so try to cope with it, provided the leak isn't too bad, until *all* leaks have been found.

When the feed-and-expansion cistern starts to fill, it is time to bleed the radiators and other vent points of any trapped air. Start at the lowest point of the system. With a radiator key, partly unscrew the radiator air vent needle – you should hear a hissing as the water forces the trapped air out. Have a cloth ready and when water starts to come out of the air vent – at anything from a dribble to a spurt – quickly screw the needle home. Continue like this round the circuit – shortly after you have finished, the cistern should stop filling up. Check the water level in the cistern and adjust the float arm if necessary.

Now connect a length of hose to the lowest drain point and allow water to run through to flush out any debris in the pipes. Close the drainvalve, attend to any leaks and remove the temporary piece of pipe and install the pump. Then refill the system, with the addition of a pre-commissioning cleanser, and bleed at the air vents again.

Starting up the boiler

Have the gas supply to the boiler connected up and tested. Once this is done, you need to set the controls and boiler thermostat to their maximum settings and, with the pump running at maximum speed,

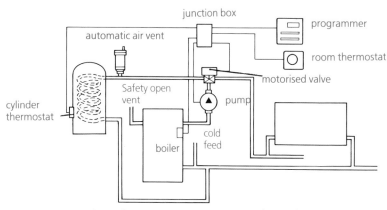

How the main components are connected together

cleanse the system according to the cleanser manufacturer's instructions.

During the cleansing process, check again for any leaks, and finally drain the system whilst hot and flush through as instructed.

> *Note: Turn off the cold supply to the feed-and-expansion cistern when draining or fresh cold water running into the still hot boiler could cause damage.*

You can now re-fill the system for the last time, adding a corrosion proofer such as Fernox – this is very important in a system containing steel panel radiators which, under certain conditions, can corrode very quickly – and vital if you have aluminium radiators.

The next step is to adjust the pump setting and balance the system, but first of all set the boiler thermostat to its maximum setting. A boiler works most efficiently at its highest temperature and the house or hot water will get up to temperature more quickly – you won't be wasting fuel since the control system thermostats will shut off the boiler when they are all satisfied. With a *condensing boiler*, however, you should use the lowest temperature that will meet your heating needs.

Set the pump speed control (if it has one) to its middle setting and check as far as you can that all controls are working properly and that hot water is not being pumped into the feed-and-expansion cistern.

The same testing and cleansing procedures apply to sealed systems, except that the system is filled via the fill unit and cleanser/inhibitor is added by removing a plug at the top of a radiator and using a funnel. Check when filling that the pressure does not rise above the expansion vessel cold-fill limit (usually ½bar).

You now have to alter the water flow through each of the radiators so that the temperature drop matches the level assumed during design –

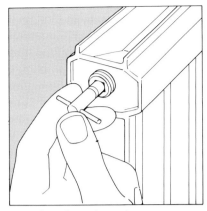

Using the radiator air vent key

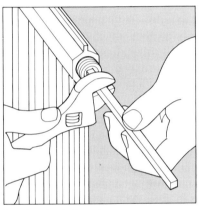

Opening the draincock

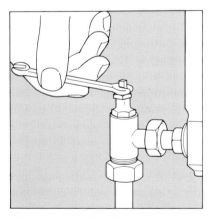

Adjusting lockshield radiator valve (when balancing)

this is called **balancing**. If you fail to get it right, the system may not achieve the temperatures it is designed for.

You alter the temperature drop by

adjusting the lockshield valve. Check the temperatures with two clip-on pipe thermometers, one on the pipework at each end of the radiator. Start with the radiator furthest away from the boiler, and check the flow temperature. If this is not 82°C, check the boiler thermostat setting and any other controls. Then gradually close the lockshield valve until the temperature drop is 11°C (i.e. the return temperature is 71°C). You may have to turn the room thermostat to its maximum setting to keep the system running. Both flow and return temperatures will tend to fluctuate as the boiler thermostat cuts in and out once the boiler is up to the temperature – it is the average temperature *difference* (11°C) that matters rather than the actual flow and return temperatures.

Repeat for all other radiators, then re-check them all. You may have to go back and re-adjust some of them and that adjustment may in turn upset the balance of other radiators: balancing can be very fiddly work.

If you are unable to get a temperature difference of 11°C you may nee a faster pump speed if the difference is too large, or a slower one if it is too small.

Insulation

The last thing to do is to insulate all the pipework that is not exposed in heated areas, including the pipework around the boiler and that to the hot water cylinder – you do not want to waste heat from this pipework during the summer. Use a good quality pipe insulation – not felt wrapping. Make sure that the insulation is firmly taped in position – electrical insulating tape is ideal for this – and, if possible, also stick the edges with a contact adhesive. Where there is a risk of freezing (such as in the roof space) stick two or more widths of insulation together to wrap around valves and any compression fittings.

Addresses

When planning plumbing work, it is helpful to have catalogues and other literature from various manufacturers – particularly if they contain design information.

The list given here includes the majority of manufacturers of plumbing goods and fittings and various central heating components (including controls). It does not include manufacturers of kitchen and bathroom equipment, nor makers of central heating boilers.

Although the list is as comprehensive as possible within the space limitations, you may also need to refer to home improvement magazines for information on new products and, of course, check in your local suppliers – particularly the d-i-y superstores (B&Q, Do It All, Great Mills, Homebase, Texas, Wickes, etc.), who have a large range of plumbing and central heating goods.

Manufacturers

Airton Industries, Unit 6, Nuffield Industrial Centre, Sandy Lane West, Littlemore, Oxfordshire OX4 5JS
☎ 0865 717611
Packaged hot water cylinders

Alumasc Ltd, Burton Latimer, Kettering, Northants NN15 5JP
☎ 0536 72 5121
Aluminium rainwater systems

Aqua-Dial Ltd, 2 Portsmouth Road, Kingston-upon-Thames, Surrey KT1 2LU
☎ 081-549 7812
Water softeners/conditioners

Aqualisa Products, The Flyers Way, Westerham, Kent TN16 1DE
☎ 0959 563240
Showers

Ballofix (Valves) Ltd, 68 Lower City Road, Tividale, Warley, West Midlands B69 2HF
☎ 021-552 5281
Ball-type isolating valves and draincocks

Barking Grohe Ltd, Brasswall House, 1 River Road, Barking, Essex IG11 0HD
☎ 081-594 7292
Taps and valves

Berglen Group Ltd, Unit 1, Kingsbury Trading Estate, Barningham Way, Kingsbury NW9 8AU
☎ 081-205 1133
Taps, valves and water filters

Bisque Ltd, 244 Belsize Road, London NW6 4BT
☎ 071-328 2225
'Styled' radiators and convectors

Brefco (UK) Ltd, PO Box 16, Brookhouse, Peel Green, Eccles, Manchester M30 7QA
☎ 061-789 8111
Vitreous enamelled and twin-walled flue systems, sealed system components, pumps

Caradon Heating Ltd, PO Box 103, National Avenue, Hull, North Humberside HU5 4JN
☎ 0482 492251
Boilers and radiators

Caradon Terrain Ltd, Aylesford, Kent ME20 7PJ
☎ 0622 717811
Plastic soil, waste, rainwater and underground drainage

Cico Chimney Linings, Westleton, Saxmundham, Suffolk IP17 3BS
☎ 0728 73608
Insulated chimney linings

Central Heating Supplies, Heating House, 134 Moss Grove, Kingswinford, West Midlands DY6 7LN
☎ 0384 278271
Mail-order suppliers of central heating equipment

Conex Sanbra Ltd, Whitehall Road, Tipton, West Midlands DY4 7JU
☎ 021-557 2831
Fittings for copper and plastic tube

Delta Capillary Products Ltd, Alexander Street, Dundee DD3 7DT
☎ 0382 21301
End-feed capillary fittings for copper pipe

Danfoss Randall Ltd, Ampthill Road, Bedford MK42 9ER
☎ 0234 364621
Electronic programmers

Dimplex Heating Ltd, Millbrook, Southampton SO9 2DP
☎ 0703 777117
Electric heaters and water heaters

Drayton Controls (Engineering) Ltd, Chantry Close, West Drayton, Middlesex UB7 7SP
☎ 0895 444012
Central heating controls

Econa Appliances Ltd, Units 15–20, Coleshill Industrial Estate, Station Road, Birmingham B46 1JR
☎ 0675 464460
Waste disposal units

Ecowater Systems, Unit 1, The Independent Business Park, Mill Road, Stokenchurch, nr High Wycombe, Bucks HP14 3TP
☎ 0494 484000
'Permutit' water softeners

Essex Partners Ltd, 73 Park Lane, Liverpool L1 5EX
☎ 051-709 6636
'Essex' flanges

Ferham Products, PO Box 164, Greasbro Road, Tinsley, Sheffield S9 1TJ
☎ 0742 446451
Cold water cisterns

Fernox Manufacturing Co Ltd, Britannica Works, Clavering, Essex CB11 4QZ
☎ 079 955 0811
Specialist domestic water treatment products and corrosion proofers

Finrad Ltd, 62 Norwood High Street, London SE27 9NW
☎ 081-670 6987
Skirting radiators

Finnish Valve Co, 279–291 Balham High Road, London SW17 7BA
☎ 081-767 4521
Taps and valves

Gledhill Water Storage Ltd, Sycamore Trading Estate, Squires Gate Lane, Blackpool, Lancs FY4 3RL
☎ 0253 401494
Hot water cylinders

Glynwed Foundries, PO Box 3, Sinclair Works, Ketley, Telford, Shropshire TF1 4AD
☎ 0952 641414
Cast iron soil and rainwater systems

Grundfos Pumps Ltd, Grovebury Road, Leighton Buzzard, Beds LU7 8TL
☎ 0525 850000
Central heating pumps

John Guest Ltd, Horton Road, West Drayton, Middlesex UB7 9PD
☎ 0895 449233
'Speedfit' plastic push-fit fittings and 'Speedpex' pipe

Harrison McCarthy Ltd, Little Moss Lane, Pendlebury, Swinton, Manchester M27 2PX
☎ 061-794 9021
Mail-order suppliers of central heating equipment

Heatrae Sadia Heating Ltd, Hurricane Way, Norwich NR6 6EA
☎ 0603 424144
Electric water heaters

Hepworth Building Products Ltd, Hazlehead, Stockbridge, Sheffield S30 5HG
☎ 0226 763561
Plastic guttering and 'Acorn' plastic pipe and fittings

Honeywell Control Systems Ltd, Honeywell House, Charles Square, Bracknell, Berks RG12 1EB
☎ 0344 424555
Central heating controls

Hunter Building Products, Nathan Way, Woolwich Industrial Estate, London SE28 0AE
☎ 081-855 9851
'Genova' plastic pipe; plastic soil, waste, underground drainage and rainwater systems

IMI Range Ltd, PO Box 1, Bridge Street, Stalybridge, Cheshire SK15 1PQ
☎ 061-338 3353
Hot water cylinders

IMI Yorkshire Copper Tube Ltd, East Lancashire Road, Kirkby, Liverpool L33 7TU
☎ 051-546 2700
Plain and plastic-covered copper tube

IMI Yorkshire Fittings, PO Box 166, Leeds LS1 1RD
☎ 0532 701104
Compression and capillary fittings for copper pipe plus valves and central heating fittings; 'Sidewinder' cylinder conversion coil

Kay & Co (Engineers) Ltd, Kontite Works, Moor Lane, Bolton BL1 4TH
☎ 0204 21041
'Kontite' fittings for copper, plastic and lead pipe

Landis & Gyr Building Control (UK) Ltd, 2 Dukes Meadow, Millboard Road, Bourne End, Bucks SL8 5XF
☎ 0628 850808
Central heating controls

Marley Extrusions Ltd, Dickley Lane, Lenham, Maidstone, Kent ME17 2DE
☎ 0622 858888
Waste, soil, underground drainage and rainwater systems

Mira Ltd, Cromwell Road, Cheltenham, Glos GL52 5EP
☎ 0242 221221
Showers and accessories

Myson Heating, Eastern Avenue, Team Valley Trading Estate, Gateshead, Tyne & Wear NE11 0PG
☎ 091-487 2211
Central heating equipment

Opella Ltd, Twyford Road, Rotherwas Industrial Estate, Hereford HR2 6JR
☎ 0432 357331
Taps, valves and d-i-y plumbing accessories

Oracstar Ltd, Weddell Way, Brackmills, Northampton NN4 0HS
☎ 0604 702181
D-i-y plumbing accessories

Pegler Ltd, St Catherines Avenue, Doncaster, South Yorkshire DN4 8DF
☎ 0302 368581
Taps, valves and central heating controls

Phetco (England) Ltd, 26a High Street, Totton, Southampton S04 4HN
☎ 0703 663777
'Multikwik' WC connectors

Polycell Products Ltd, Broadwater Road, Welwyn Garden City, Herts AL7 3AZ
☎ 0707 328131
Plastic push-fit fittings and d-i-y plumbing kits

Polypipe plc, Broomhouse Lane, Edlington, Doncaster DN12 1ES
☎ 0709 770000
Plastic waste, soil and rainwater systems

Potterton Myson Ltd, Portobello Works, Emscote Road, Warwick CV34 5QU
☎ 0926 493420
Central heating equipment

Ravensbourne Heating Ltd, 34–50 Cemetery Road, Lye, Worcs DY9 7EQ
☎ 0384 423841
Mail-order suppliers of central heating equipment

Redring Electric, Celta Road, Woodstone, Peterborough PE2 9JJ
☎ 0733 313213
Electric heaters, water heaters and showers

Robimatic plc, Unit 2, Red Kiln Close, Horsham RH13 5TP
☎ 0403 67528
Self-cutting valves and d-i-y plumbing accessories

Runtalrad (1970) Ltd, The Ridgeway Road, Iver, Bucks SL0 9JQ
☎ 0753 655215
'Styled' radiators

Salamander (Engineering) Ltd, Reddicap Trading Estate, Sutton Coldfield, West Midlands B75 7BY
☎ 021-378 0952
'Scalemaster' scale reducer

Saniflo Ltd, Howard House, The Runway (off Station Approach), South Ruislip, Middx HA4 6SE
☎ 081-842 0033
Pumping-and-shredding WC units

Selkirk Manufacturing Ltd, Bassett House, High Street, Banstead, Surrey SM7 2LZ
☎ 073 73 53388
Insulated block chimneys

Stanton plc, PO Box 72, Nottingham NG10 5AA
☎ 0602 322121
Cast iron soil and drainage systems

Tanks & Drums Ltd, Bowling Iron Works, Bradford, West Yorks BD4 8SX
☎ 0274 728285
Cold water cisterns and unvented hot water systems

Triton plc, Newdegate Street, Nuneaton, Warwicks CV11 4EU
☎ 0203 344441
Electric showers and shower booster pump

Tufcon Building Products Division, Units 33/34, Alexandra Way, Ashchurch Business Centre, Tewkesbury, Glos GL20 8NB
☎ 0684 850915
Insulated pipe ducting and pipe hiding systems

Wavin Building Products Ltd, Parsonage Way, Chippenham, Wilts SN15 5PN
☎ 0249 654121
'Osma' plastic soil, waste, rainwater and underground drainage systems

Wednesbury Tube, Oxford Street, Bilston, West Midlands WV14 7DS
☎ 0902 491133
Copper tubes and fittings

Other useful addresses

Chartered Institution of Building Services Engineers (CIBSE), Delta House, 222 Balham High Road, London SW12 9BS
☎ 081-675 5211

Council for Registered Gas Installers (CORGI), 4 Elmwood, Chineham Business Park, Crockford Lane, Basingstoke, Hants RG24 0WG
☎ 0256 707060

Electrical Contractors Association (ECA), ESCA House, 34 Palace Court, Bayswater, London W2 4HY
☎ 071-229 1266

Heating and Ventilating Contractors Association (HVCA), ESCA House, 34 Palace Court Road, Bayswater, London W2 4JG
☎ 071-229 2488

Institute of Plumbing, 64 Station Lane, Hornchurch, Essex RM12 6NB
☎ 040 24 72791

National Association of Plumbing, Heating and Mechanical Services Contractors (NAPHMSC), Ensign House, Ensign Business Centre, Westwood Way, Coventry CV4 8JA
☎ 0203 470626

National Inspection Council for Electrical Installation Contracting (NICEIC), Vintage House, 37 Albert Embankment, London SE1 7UJ
☎ 071-735 1322

Solar Trade Association Ltd, Brackenhurst, Greenham Common South, Newbury, Berks RG15 8HH
☎ 0635 46561

Index